Regenerative Dentistry

Regenerative Dentistry

Mona K. Marei, Editor

ISBN: 978-3-031-01453-6 paperback
ISBN: 978-3-031-02581-5   ebook

DOI 10.1007/978-3-031-02581-5

A Publication in the Morgan & Claypool Publishers series
SYNTHESIS LECTURES ON TISSUE ENGINEERING

Lecture #6
Series Editor: Kyriacos A. Athanasiou, *University of California, Davis*
Series ISSN
Synthesis Lectures on Tissue Engineering
Print 1944-0316   Electronic 1944-0308

Synthesis Lectures on Tissue Engineering

Editor
Kyriacos A. Athanasiou, *University of California, Davis*

The Synthesis Lectures on Tissue Engineering series will publish concise books on aspects of a field that holds so much promise for providing solutions to some of the most difficult problems of tissue repair, healing, and regeneration. The field of Tissue Engineering straddles biology, medicine, and engineering, and it is this multi-disciplinary nature that is bound to revolutionize treatment of a plethora of tissue and organ problems. Central to Tissue Engineering is the use of living cells with a variety of biochemical or biophysical stimuli to alter or maximize cellular functions and responses. However, in addition to its therapeutic potentials, this field is making significant strides in providing diagnostic tools. Each book in the Series will be a self-contained treatise on one subject, authored by leading experts. Books will be approximately 65-125 pages. Topics will include 1) Tissue Engineering knowledge on particular tissues or organs (e.g., articular cartilage, liver), but also on 2) methodologies and protocols, as well as 3) the main actors in Tissue Engineering paradigms, such as cells, biomolecules, biomaterials, biomechanics, and engineering design. This Series is intended to be the first comprehensive series of books in this exciting area.

Regenerative Dentistry
Mona K. Marei, Editor
2010

Cells and Biomaterials for Intervertebral Disc Regeneration
Sibylle Grad, Mauro Alini, David Eglin, Daisuke Sakai, Joji Mochida, Sunil Mahor, Estelle Collin, Biraja Dash, Abhay Pandit
2010

Fundamental Biomechanics in Bone Tissue Engineering
X. Wang, J.S. Nyman, X. Dong, H. Leng, M. Reyes
2010

Articular Cartilage Tissue Engineering
Kyriacos A. Athanasiou, Eric M. Darling, Jerry C. Hu
2009

Tissue Engineering of Temporomandibular Joint Cartilage
Kyriacos A. Athanasiou, Alejandro J. Almarza, Michael S. Detamore, Kerem N. Kalpakci
2009

Engineering the Knee Meniscus
Kyriacos A. Athanasiou, Johannah Sanchez-Adams
2009

Regenerative Dentistry

Mona K. Marei, Editor
Alexandria University, Egypt

SYNTHESIS LECTURES ON TISSUE ENGINEERING #6

ABSTRACT

Dental caries, periodontitis, tooth loss, and bone resorption are considered prevalent health problems that have direct affect on the quality of life. While, advances in stem cell biology and biotechnology have sparked hope for devastating maladies, such as diabetes, cardiovascular diseases ..., etc., it also provides a strategy of regenerative therapy for dental tissues. From the prospective of tissue engineering, it is of utmost importance to understand and emulate the complex cell interactions that make up a tissue or organ. Unlike other tissues in the body, dental tissues are unique in their development, function, and even in their maintenance throughout life.

The harmonized stimulations of biology and mechanical regulators to promote cellular activities have matured our understanding of the value of regenerative therapy of dental tissue versus the reparative treatment.

In this book, we review the current knowledge available to regenerate alveolar bone, periodontal structure, and pulp/dentin complex.

The book provides researchers with detailed information about development and functional characteristics of the dental unit with detailed protocols covering a comprehensive range of various approaches to engineer dental tissues: to use isolated cells or cell substitutes as cellular replacement, to use acellular biomaterials capable of inducing tissue regeneration, and/or to use a combination of cells, biomaterial and growth factors. We are well aware, with the concept changes in the field toward in-vitro biomimetics of in-vivo tissue development. The theoretical frame work integrating these concepts of developmental biology and developmental engineering is yet to be emphasized and implemented.

Until this happens, we consider this book of regenerative dentistry as a call for scientists to achieve, researchers to innovate, practitioners to apply, and students to learn the art and science of regenerative therapy in dentistry.

KEYWORDS

biomaterials in dental tissue regeneration, alveolar extraction socket, tissue engineering alveolar bone, biomimetic scaffolds, osteogenic proteins, cementogenesis, bone induction, dental pulp stem cells, dentinogenesis, tertiary dentinogenesis

To my daughters;
Rania & Reem.

Contents

Rania M. Elbackly, BDS, MSc
Manal M. Saad, BDS, PhD
Ahmad Rashad, BDS, MSc
Samer H. Zaky, BDS, PhD

3   **Tissue Engineering of the Periodontal Tissues** 83

Ugo Ripamonti, MD, PhD
Jean-Claude Petit, BSc, LDS, MDent
June Teare, MSc

Preface

Advances in stem cell biology and biomaterials development in the late 1990s have helped drive on an ever expanding body of research in the field of tissue engineering and regenerative medicine. Scientists realized that the key to future success of functional tissues is bridging the gap between developmental biology and tissue engineering. We are all amazed by the high degree of sophistication and miniaturization found in nature. Nature is, indeed, a school of science.

While tissue engineering and regenerative medicine inspired research and applications in areas like bone, cartilage and skin, it challenges the building of completely artificial organs, e.g., liver. The urgency of clinical emergencies drove the approaches toward all living tissues in the body, including dental structures. There is a staggering dental need for new and effective treatment for acquired as well as inherited defects of dental tissues. This monograph on regenerative dentistry provides researchers and clinicians with updates on the science and applications of regeneration of dental tissues. The introduction throws light on the field from the standpoint of material and biological key players that govern the current paradigms of regenerative dentistry. The following 3 chapters are largely the product of expertise in three main dental tissues: regeneration of alveolar bone, regenerative periodontics, and regenerative endodontics. These chapters focus on the current knowledge that has revolutionized the regeneration of these dental tissues and paves the way to the hope of strategic direction in the science and technology of regenerative dentistry. We all appreciate the value of discussing the fundamentals of regeneration of each of these tissues and the potential challenges offered by current publications that open the venue to future researches and tools in these areas.

Mona K. Marei, Editor
July 2010

CHAPTER 1

Introduction to Regenerative Dentistry

Charles Sfeir, DDS, PhD

Shinsuke Onishi, DDS, PhD

Sayuri Yoshizawa, DDS, PhD

Fatima Syed-Picard, MS

1.1   CHAPTER SUMMARY

Humans, unlike species such as salamander or newt, lack the ability to naturally regenerate their own tissues. To overcome this limitation, tissue engineering strategies utilizing combinations of biocompatible scaffolds, growth factors, and stem cells to mimic natural morphogenesis, are currently in development. In humans, morphogenesis during embryological development relies on the spatial organization of cells to give rise to tissues and organs. While these morphogenetic events are under genetic control, the formation of specialized structures is governed by the microenvironment in which the cells reside. Tissue engineering aims to emulate morphogenesis for the purpose of tissue and organ regeneration, both in the laboratory as well as *in situ*. The overall goal of these strategies is to provide the delivered (as well as surrounding) cells with an environment in which they can proliferate, differentiate, and mature to form the desired structure. Several research groups have been inspired by the natural architecture and composition of specific organs. This inspiration has led to the design of biomimetic scaffolds aimed at achieving an efficient regenerative process. In addition, novel tissue engineering methodologies such as bio-printing have been used to better control the delivery of growth factors and cells in a three-dimensional environment. These new methodologies aim to leverage developmental biology principles toward the engineering of organs. Such efforts may prove critical for *in vitro* organ building.

In this chapter, we will summarize the various tissue engineering strategies utilized in dentistry. We will cover the three main components necessary to successfully engineer a tissue or organ, namely materials, growth factors, and cells, and explain how these components vary among the different dental tissue structures.

1.2 MATERIALS USED IN DENTAL TISSUE REGENERATION

The biomaterial component of tissue regeneration involves engineering scaffolds for the creation of three-dimensional tissue structures. These scaffolds must exhibit biomimetic properties by which they simulate the advantageous features of the natural tissue microenvironment. Effective scaffolds can facilitate cell recruiting/seeding, adhesion, proliferation, differentiation and tissue neogenesis (Ma, P., 2008). In addition, these materials must be able to endure the mechanical environment in which they will be placed and have the ability to be eventually safely degraded and replaced by natural tissue. Three different categories of materials exist: polymers, ceramics, and metals. The varying properties of these materials make them suitable scaffold materials for the engineering of different types of craniofacial tissues.

Due to the ease of isolation and similarities with soft tissues, natural and synthetic polymers are often used as tissue engineering scaffolds. Natural polymer scaffolds include collagen (Cen et al., 2008; Rohanizadeh et al., 2008), hyaluronic acids (Burdick et al., 2005), calcium alginate (Cho et al., 2005), and chitosan (Kim et al., 2008). Synthetic polymer scaffolds include polyglycolic acid (PGA), polylactic acid (PLA) (Athanasiou et al., 1996), Poly-LD-lactic acid/polyglycolic acid (PLGA) (Tschon et al., 2009), and hydrogels (Trojani et al., 2006).

Ceramic scaffolds are used to mimic the hydroxyapatite found in mineralized tissue. Specifically, calcium phosphate ceramics such as hydroxyapatite (Mankani et al., 2008) and beta-tricalcium phosphate (Kamitakahara et al., 2008) are often used in bone tissue engineering applications. Also, due to its biocompatibility and similar mechanical properties to bone, bioactive glass can be utilized as a scaffold material (Day et al., 2005).

To date, there has been only limited research involving the use of metals as tissue engineering scaffolds. Nevertheless, metals such as titanium, iron alloys and stainless steel are commonly used in the dental implant field. In addition, more recently, biodegradable metal scaffolds such as magnesium alloys, have been assessed for bone regenerative potential (Witte et al., 2007).

In addition to these three types of traditional materials, self-assembled peptides (Ulijn and Smith, 2008) have also been used as scaffolds. Certain combinations of natural amino acids can be self-assembled to form specific structures, such as beta-sheets, alpha-helices, and peptide amphiphiles. These structures can be modified with bioactive molecules to mimic natural extracellular matrices.

In the next section, we will examine the materials used for tissue engineering applications of specific craniofacial tissues.

1.2.1 MATERIALS USED FOR TOOTH REGENERATION

Various biodegradable polymer scaffolds have been used for tooth regeneration. These polymers include poly-L-lactate-co-glycolate (PLGA) (Young et al., 2002), silk (Xu et al., 2008), gelatin-chondroitin hyaluronan (Kuo et al., 2008), collagen gels or sponges (Honda et al., 2007; Ikeda et al., 2009), and self-assembled peptides (Kirkham et al., 2007).

A study by Young et al., was recently performed in which PLGA was shaped into crowns of human incisors and molars, and coated with collagen type I gel. The total cell population of porcine third molar tooth buds was seeded into the PLGA and the constructs were transplanted into the omentum of athymic rats. After 20 to 30 weeks, a crown-like enamel and dentin structures were observed (Young et al., 2002).

In another type of approach, silk scaffold material was fabricated into porous sponges via salt porogen methods. Rat tooth bud cells were harvested, directly mixed with Matrigel, seeded onto the porous silk scaffold, and transplanted into syngeneic Lewis rats. The cells formed small mineralized particles distributed in the region of the scaffold that degraded most after 20 weeks (Xu et al., 2008).

The above referenced gelatin-chondroitin-hyaluronan tri- copolymer scaffold was prepared by cross-linking using 2mL of 1% 1-ethyl-3-(3-dime- thylaminopropyl) carbodiimide (EDAC). This scaffold has 75% porosity. Porcine molar bud cells were cultured and transplanted with the scaffold into the original alveolar socket of donor swine. The transplants were harvested at 36 weeks, and small $(0.5 - 1 \times 0.5 \times 0.5 \ cm^3)$ but well-organized enamel and dentin tissue were formed (Kuo et al., 2008).

In a fourth type of approach, beta-sheet-forming peptides were designed as a biomimetic scaffold for non-cellular tissue engineering of enamel remineralization. The scaffold spontaneously forms a three-dimensional fibrillar form in response to specific environmental triggers. This technique may potentially be used not only in the treatment/prevention of dental caries but also in the modification of the surface of other scaffolds (Kirkham et al., 2007).

In a study by Honda et al., the authors created a collagen sponge consisting of 75% (dry weight) type I atelocollagen and 25% type III atelocollagen from porcine skin. The sponge was transplanted into porcine mandible with odontogenic mesenchymal and epithelial cells derived from porcine third molar buds. After 4 to 8 weeks, enamel and dentin-like tissues were observed in the transplanted groups where mesenchymal cells were plated within the collagen sponge and epithelial cells were plated on top (Honda et al., 2007). In a more recent study by Ikeda et al., whole functional teeth were regenerated through the use of collagen gel reconstituted *ex vivo* with highly concentrated mesenchymal and epithelial cells of mouse tooth bud. Following culture of the construct for 5-7 days, the reconstituted tooth germ was transplanted into adult mouse into a socket from which a tooth had been extracted. The whole tooth was regenerated in the socket. The cusp tip of the bioengineered tooth erupted spontaneously to reach the plane of occlusion with the opposing lower first molar, at 49.2 ± 5.5 days after transplantation. This study indicates the possibility of tooth replacement therapy. However, further studies to identify available adult stem cells for the reconstitution of a bioengineered tooth germ and the regulation of stem cell differentiation into odontogenic cell lineage are needed (Ikeda et al., 2009).

Ceramic scaffolds made from hydroxyapatite/tricalcium phosphate (HA/TCP) ceramic powder and used along with adult human dental pulp stem cells (DPSCs), are reported to form dentin and pulp-like structures when transplanted subcutaneously into immunocompromised mice (Gronthos et al., 2000). Hydroxyapatite (HA) is an analogue of the mineral component of the bone tissue, and it is stable against dissolution by body fluids, whereas tricalcium phosphate (TCP) has a much higher resorption rate compared to that of HA. By varying the component ratio in an HA/TCP composite, one can control the resorption rate (Kubarev et al., 2007).

1.2.2   BONE

Over the last two decades, a large effort has been directed toward bone reconstruction. Recently, INFUSE®, a biodegradable collagen sponge scaffold coated with bone morphogenetic protein-2 (BMP-2), was approved for clinical treatment of bone defects. BMP-2, a member of the transforming growth factor-beta (TGFβ) cytokine super family, is a multifunctional growth factor known to enhance bone regeneration. INFUSE® constructs containing a combination of this cytokine and the collagen scaffold can be completely absorbed and replaced by regenerated bone.

It has also been reported that a combination of HA/TCP particles with bone marrow stromal cells (BMSC) created a complex containing new bone and bone marrow (Mankani et al., 2008). Mankani et al., mixed 0.5-1mM HA/TCP particles with human bone marrow stromal cells and transplanted them subcutaneously in the backs, or subperiosteally on the calvarium, of immunocompromised mice (Mankani et al., 2008). After 7-110 weeks, the construct formed extensive bone and bone marrow, and HA/TCP was partially degraded, but the HA/TCP did not degrade completely even after 107 weeks. This study shows that HA/TCP is a promising material for bone regeneration, however, its degradation properties need to be tailored.

Since metals are more suitable for load-bearing applications due to their combination of strength, ductility, and toughness, metal alloys have been analyzed as potential bone scaffolds. Witte et al. (2007) studied magnesium alloy such as AZ91D (9 wt % aluminium, 1 wt % zinc, 0.15-0.5 wt % manganese in pure magnesium), due to its biodegradation capabilities. This alloy was cast in a negative salt-pattern molding process to create porous scaffolds. The salt particles were then washed out resulting in a porous scaffold. These scaffolds were transplanted into distal femur condyle of rabbits. The magnesium alloys did not cause significant harm to their neighboring tissues according to immunohistochemical analysis. Transplanted magnesium alloys were degraded after three months, and although the tissue surrounding the transplant showed increased bone deposition and remodeling, the alloy itself was replaced with fibrous tissue and only a limited number of cells (Witte et al., 2007). A greater amount of trabecular bone, more osteoblasts, and thicker osteoid were observed around the magnesium alloy scaffold as compared to the transplanted autologous bone. These results indicate that magnesium alloys are biocompatible and promote bone remodeling around the scaffold.

Overall, the different materials currently used for bone engineering show promising preliminary results, but these materials need to be further tailored for large bone defect therapies. HA/TCP induced highly differentiated tissue regeneration, but the mechanical properties of the scaffold is not enough to support the load-bearing conditions. Magnesium alloys have suitable mechanical properties, but good osteoinductivity has not been observed. Modification of mechanical properties and osteoinductivity will be key to the development of bone regenerative scaffolds. Additionally, in the case of the collagen sponge, mouse (but not human) BMSC were induced to form bone (Krebsbach et al., 1997), stressing the fact that potential regenerative therapies must be assessed specifically using human cells.

1.2.3   TEMPORO-MANDIBULAR JOINT (TMJ)

The TMJ is a multi-tissue structure composed of articular cartilage disc connected to subchondral bone. Due to its complexity, the TMJ is a difficult structure to engineer, however, many attempts have been made. Engineering a functional osteochondral graft requires the production of both bone and cartilage with a defined interface (Scheller et al., 2009), along with a scaffold that should be matched both biologically and functionally to the host defect. Schek et al. (2005) designed a 3D scaffold for mandibular condyle regeneration where the scaffold shape and internal pore architecture were defined by images of MRI and/or CT. The scaffolds used were composites of hydroxyapatite and PLA. These were manufactured using 3D inkjet printing and a salt leaching technique. The scaffolds were implanted subcutaneously with pig cartilage cells infected with an adenovirus coding for BMP7 (AdCMV-BMP-7). After 4 weeks, the cells formed a small amount of bone and cartilage-like tissue.

Alhadlaq et al., demonstrated the successful regeneration of the condyle using photopolymerized polyethylene (glycol) diacrylate (PEGDA) (Alhadlaq and Mao, 2003). This was achieved using rat bone-marrow mesenchymal stem cells which had been treated with chondrogenic or osteogenic differentiation factors and seeded in PEGDA solution containing a biocompatible ultraviolet photoinitiator. PEGDA solution with chondrogenic cells was loaded into the cartilage region of a human mandibular condyle-shaped polyurethane mold and cured by UV light, and a PEGDA solution of osteogenic cells was filled in the remaining bone region of the mold. The construct was implanted subcutaneously into immunocompromised mice, and 4 weeks after transplantation, the constructs were shown to contain both chondrogenic and osteogenic layers.

These scaffolds seem to have potential toward regenerating each tissue component of the TMJ. However, the studies were not performed with human cells and the *in vivo* studies were only performed subcutaneously. In order to achieve the regeneration of a functional TMJ, there is a strong need for further research in this area.

1.3   PROTEIN, GENE, AND OTHER MOLECULE DELIVERY SYSTEMS

The targeted delivery of regenerative factors is one of the strategies by which we try to stimulate naturally occurring processes in mammalian tissue development and repair. The local administration of specific bioactive molecules or signals may induce tissue formation by enhancing the progenitor cells to migrate from adjacent sites to the repair site and/or by accelerating differentiation and proliferation processes (Alsberg et al., 2001). This can be accomplished by several different approaches, including the direct delivery of recombinant protein to the target site, exogenous gene delivery aiming to sustain bioactive signals over long periods of time, and direct bioactive mediator/drug delivery.

Among such deliverable signals, the most frequently used regenerative factors are growth factors and their genes, which play pivotal roles in natural tissue remodeling and have been shown to have an inductive effect on tissue formation during repair processes (Baum and Mooney, 2000).

Studies have shown that the systemic and local injection of regenerative factors can effect new bone formation in animal models (Kato et al., 1998; Chen et al., 2003; Einhorn et al., 2003). It has been argued that the relatively rapid diffusion of these factors, due to lack of carriers to contain them to the target site, may reduce the efficacy of the factors in promoting new bone formation. These factors are therefore often delivered through matrices or scaffolds that, in addition to exhibiting the required properties of scaffold materials mentioned earlier, can also retain these factors for a prolonged time. Additionally, the best systems also deliver the factor to the targeted site in a controlled manner so that it is released at the appropriate time. The use of materials which have the specific features mentioned above, can achieve more effective and rapid tissue repair than those which do not. Combinations of delivery systems and regenerative factors are highlighted in Table 1.1.

Table 1.1: Summary of delivery systems for growth factors and other regenerative factors involved in dentin, pulp, periodontium, and temporomandibular joint regeneration.

Tissue type	Growth factor	Carrier	References
Dentin and pulp	FGF-2	Gelatin hydrogels	(Ishimatsu et al., 2009)
	DMP-1	Collagen	(Prescott et al., 2008)
	BMP-2	Collagen sponge	(Nakashima, M., 1994)
Periodontium	PDGF and Dexamethasone	Collagen and Gelatin hydrogels	(Rutherford et al., 1993)
	BMP-2 and IGF-1	Gelatin microparticles and Hydrogels	(Chen et al., 2009)
Temporomandibular joint	FGF-2	Collagen	(Takafuji et al., 2007)
	BMP-2	Polylactic acid/ polyglycolic acid copolymer and gelatin sponge complex	(Ueki et al., 2003)

Abbreviations: FGF-2, fibroblast growth factor-2; DMP-1, dentin matrix protein-1; BMP-2, bone morphogenetic protein-2; PDGF, platelet-derived growth factor; IGF-1, insulin growth factor-1.

1.3.1   PROTEIN DELIVERY

Protein delivery for tissue engineering describes the targeting of single or multiple proteins or peptides to a specific site. The goal of such delivery is the promotion of new tissue formation or enhancement of naturally-occurring tissue repair by recruiting or activating neighboring progenitor cells. Proteins or peptides utilized for this purpose include growth factors, extracellular matrix (ECM) proteins and attachment molecules (Cochran and Wozney, 1999). As mentioned before, these proteins are involved in tissue formation at varied stages and timings leading to the enhanced response of adjacent target cells. For instance, growth factors play essential roles in tissue formation such as cell proliferation, differentiation, maturation, vascularization, and ECM synthesis. Fibronectin, one of the ECM proteins related to cell binding, has been shown to mediate the binding of signaling molecules for the reorganization of pre-odontoblast cytoskeleton in the process of pulpal wound repair (Nakashima and Akamine, 2005). Of these proteins, growth factors are probably the most frequently used for tissue regeneration.

Although a considerable interest in using growth factors for tissue engineering has been generated during the past few years, there have indeed been certain limitations. In general, most proteins have short half-lives *in vivo,* therefore premature degradation of proteins is a major obstacle. In order to compensate for this, several strategies are used for retention of proteins to their vehicles, thereby maximizing their effects on target cells. Use of biodegradable microspheres (e.g., PLGA or chitosan microspheres), in which the regenerative proteins are encapsulated, is one method that has demonstrated controlled release of regenerative factors (Lu et al., 2001; DeFail et al., 2006; Moioli et al., 2006; Cai et al., 2007). These microspheres can be injected or combined with scaffolds in order to place them at the target site. The combination of these microspheres with a three-dimensional (3-D) porous scaffold (e.g., collagen sponges, poly-L-lactic acid) (PLLA) scaffolds) provides a vehicle with better control of delivery to invading cells (Wei et al., 2006; Takemoto et al., 2008). Although these studies have successfully demonstrated significant bone formation over the controls, it may be necessary for supraphysiological dosages of those proteins to be utilized due to their short half-lives *in vivo.* Concomitantly, patient safety and cost performance issues need to be addressed.

1.3.2   GENE DELIVERY

Due to the inherent limitations of protein delivery systems, a clear need exists for gene therapy approaches in tissue engineering. One of the major advantages of gene delivery over protein delivery is that it can maintain physiological levels of the regenerative factors for prolonged periods of time at the target site. It is difficult to achieve this through protein delivery, due to the short half-life of protein.

Gene delivery methods are frequently categorized into non-viral or viral gene delivery. Although each method has its unique limitations as well as inherent advantages, the benefits of non-viral gene delivery outweigh the advantages of viral gene delivery. In general, non-viral gene delivery is less expensive, simpler, easier to use, safer, has low immunogenicity, and no size limitation to the

DNA insert. There is, however, a lower gene transfer efficiency (Scheller and Krebsbach, 2009) when compared to viral gene delivery.

Non-viral gene delivery systems can be generally grouped under the materials they utilize, such as chemicals, synthetic polymers, and natural polymers. Calcium phosphate and cationic liposomes have been used for a long time due to their simplicity of use and reasonable transfection efficiency. Use of cationic liposomes has become the widely preferred approach, due to higher transfection efficiency and the ability of transfecting a larger number of cell types.

The rationale behind synthetic polymers, is that cationic polymers become associated with plasmid DNA, which is negatively charged, resulting in a positively charged polymer/DNA complex. This complex is then internalized into a cell by endosomal uptake through interaction between the cell surface glycocalyx and the complex (Scheller and Krebsbach, 2009). To generate the cationic polymer/DNA complex and minimize its drawbacks (e.g., cytotoxicity, aggregation, low transfection efficiency), polyethylene glycol (PEG) has been used with other polymers such as poly L-lysine (PLL) (Choi et al., 1998), polyethyleneimine (PEI) (Boussif et al., 1995).

The natural polymers utilized for gene delivery contains cyclodextrin, chitosan, collagen, gelatin, and alginate. The advantages of natural over synthetic polymers include degradation ability, remodeling by cell-secreted enzymes (Scheller et al., 2009), and good cytocompatibility (Dang and Leong, 2006). Although the transfection efficiency of natural polymer is similar to that of both synthetic polymer (PEI) and artificial cationic liposomes (lipofectamine), it is significantly lower than the efficiency experienced with virus-based gene delivery methods (Gonzalez et al., 1999).

In terms of actual transfer efficiency, viral gene delivery is the most efficient method. Viral vectors achieve their success by inserting their genetic material into the genome of target cells. There are three main methods grouped under categories of their viral types: adenoviral, adeno-associated viral (AAV), and retroviral vectors. Adenoviruses are double-stranded DNA viruses with linear genomes. Their advantages include their ability to transducer a wide range of cells, both dividing and non-dividing (Verma and Somia, 1997), and they exhibit almost no stable integration into the host genome (Tenenbaum et al., 2003). Although this last advantage can diminish the risk of host cell mutagenesis, the lack of integration means that there is no replication of the viral genome, and therefore only transient transgene expression. AAV is a single-stranded DNA virus featuring a linear genome. Its advantages include good transduction into a number of different cell types, the ability to transduce both dividing and non-dividing cells, prolonged transgene expression, and episomal retention for safety. Drawbacks of AAV include incomplete elimination of the helper virus upon reproduction, and relatively low packaging capacity. Retroviruses are RNA viruses, which reverse-transcribe their RNA into DNA. Their advantages include low immunogenicity and prolonged transgene expression due to ready integration into the host genome. Disadvantages include the inability to transduce non-dividing cells (except in the case of lentivirus, which can transduce both dividing and non-dividing) and a relatively low packaging capacity.

1.3.3   DELIVERY OF OTHER MOLECULES

In addition to proteins (mainly growth factors) and genes used for craniofacial tissue engineering, therapeutic strategies may also call for the delivery of metabolism mediators for bone remodeling, such as eicosanoids and glucocorticoids. For example, prostaglandins have been shown to have multiple effects on bone resorption and remodeling (Miller and Marks, 1994). It has also been demonstrated that prostaglandins enhance new bone formation *in vivo* and *in vitro* (Marks and Miller, 1994; Damrongsri et al., 2006). Some hormones such as parathyroid hormone (PTH) and dexamethasone have been shown to have effects on bone metabolism. In a study of systematic PTH delivery in rats, an increase of bone mineral density and bone mineral content resulted, indicating that this approach could be useful for tissue engineering (Chen et al., 2003). The delivery of these mediators has potential to enhance the process of bone repair or bone formation at the target site.

1.4   CELLULAR THERAPIES FOR DENTAL TISSUE ENGINEERING

Current techniques for restoring dental or craniofacial defects include the use of synthetic materials, autografts, or allografts. These therapies, however, have limitations, such as limited tissue regeneration, lack of graft availability, or immune rejection (Scheller et al., 2009). One particularly powerful therapeutic angle on restoration looks to the body's own natural methods of repair, namely so-called stem cells, that remain quiescent in the body until needed for growth, renewal, and maintenance of tissues. Stem cells can be derived from many different sources, with the most valuable source being from blastocyte stage embryos. These pleuripotent cells have the potential to differentiate down a variety of lineages. Embryonic stem cells are involved in initial organ formation during development. They have the greatest long-term proliferation capacity and can transform into any fully differentiated cell in the body. The therapeutic possibilities for the use of embryonic stem cells are limited by ethical concerns associated with their isolation. Adult stem cells, on the other hand, maintain and repair tissues and can be found in several different regions of the body such as bone marrow, adipose, skin, blood, and muscle tissues. These cells do not have the same level of plasticity as embryonic stem cells but can still differentiate into several different cell types.

The *in situ* microenvironment of stem cells, also known as the stem cell niche, is thought to be composed of other cells and matrix that maintain the stem cells in a quiescent state. Disruptions of this niche due to such things as trauma may trigger the stem cell to multiply or differentiate, with the resulting progenitor or differentiated cells aiding in tissue repair and remodeling. A more fundamental understanding of the factors that stimulate stem cell differentiation down various pathways, particularly the formation of dental tissues from stem cells, will lead to cellular therapies and tissue engineering approaches for dental tissue regeneration. To date, stem cells have been discovered and characterized in several different adult dental tissues such as dental pulp, periodontal ligament, salivary glands, and bone marrow (Watt and Hogan, 2000; Moore and Lemischka, 2006).

1.4.1   STEM CELL ISOLATION

When tissue is harvested for stem cell isolation, the tissue is first digested and all the cells are subsequently collected. Only a small fraction of the total cells that are isolated are the actual stem cells. Several studies have attempted to culture only the stem cells from this miled population by plating the cells at a very low density. Since stem cells are clonogenic, they are able to survive and divide at very low density, whereas differentiated cells are not. These clonogenic cells eventually form individual colonies, and if cultured properly, these groups of cells will maintain their stem cell qualities. Stem cells can also be characterized by cell surface markers and different cell sorting techniques, such as fluorescence-activated cell sorting (FACS) or magnetic activated cell sorting (MACS), have been developed for the separation of cells based on expression of these extracellular markers.

To utilize FACS, the proteins of interest are tagged with fluorescently labeled antibodies, the cells are passed through a flow cytometer, and their relative fluorescence is excited using a laser light source. The cells can be characterized and separated by size, volume, and protein expression. The MACS system uses a similar method to separate cell populations. This sorting technique utilizes magnetic beads that are coated with specific antibodies. The cells are incubated with the beads and those expressing the protein of interest become attached. The beads are passed through a column generating a magnetic field, and the cells expressing the marker adhere to the column. The remaining cell populations flow through and are discarded. At this time, MACS is a bench-top apparatus that can be easily set up in most labs; however, when compared with FACs, MACS is significantly limited in its flexibility. FACS requires a flow cytometry system, which is both large and expensive. It is, however, significantly more powerful than MACS in that it is designed to characterize and isolate multiple subpopulations of cells via antigenic recognition whereas MACS only sorts the cells.

Studies have shown that mesenchymal stem cells reside in the microvascular regions of the tissues from which they are isolated. Therefore, cell surface markers unique to specific types of endothelial cells or pericytes are often used to isolate stem cell populations. At this time, discovery research of specific stem cell markers is still in its early stages. Table 1.2 lists a few of the commonly used cell surface markers and the types of cells with which they are currently believed to be associated. Finally, it is important to note that differentiated cells also play a role in tissue regeneration, and more work is required toward designing therapeutic strategies involving total cellular populations.

1.4.2   DENTAL PULP STEM CELLS

The dental pulp is the vascular connective tissue found on the inner most region of the tooth, where it is surrounded by dentin. This tissue contains connective tissue, fibroblasts, neural fibers, blood vessels and lymphatics. Recently, mesenchymal stem cells have been discovered within dental pulp tissue. These cells have been isolated and characterized from both adult and deciduous teeth (Gronthos et al., 2000; Miura et al., 2003). Both adult dental pulp stem cells (DPSC) from third molars, and stem cells from human exfoliated deciduous teeth (SHED), are clonogenic and capable of differentiating into multiple cell types *in vitro*. Their potential to date includes odon-

Table 1.2: List of common markers used to sort stem cells and the cell type to which they are associated.

Marker	Cell
STRO-1	Pericytes that are stromal cell precursors
C-kit	Hematopoietic stem cells
CD34	Stromal cell precursors
CD146	Pericytes
von Willebrand Factor	Endothelial cells
3G5	Pericytes

toblast, fibroblast, adipocyte and neural-like lineages. SHED have greater proliferation rates than adult DPSC; *in vivo* studies where the two types of stem cells were subcutaneously implanted into immunocompromised mice with a hydroxyapatite carrier revealed that adult DPSC form dentin-pulp complexes, whereas SHED form dentin and bone. Both populations of cells are clonogenic and can form multiple tissue types *in vitro*. Subcutaneous implants of adult DPSC sorted for a CD146$^+$ population have been shown to form dentin and pulp complexes, whereas similar transplantation studies utilizing c-kit $^+$/CD34$^+$/CD45$^-$ showed the formation of bone-like tissue. These studies revealed that multiple types of stem cells can be isolated from the same tissue (Shi and Gronthos, 2003; Laino et al., 2005).

1.4.3   PERIODONTAL LIGAMENT STEM CELLS

The periodontal ligament (PDL) holds the tooth in its socket by connecting the cementum to alveolar bone. This fibrous, vascular tissue contains several cell types including fibroblasts, myofibroblasts, endothelial cells, nerve cells, epithelial cells, osteoblasts, and cementoblasts. The PDL also contains a population of mesenchymal stem cells that are clonogenic and can differentiate into cementoblast-like, adipocyte-like, and fibroblast-like cells. When periodontal ligament stem cells (PDLSC) are implanted subcutaneously into immunocompromised mice with a hydroxyapatite carrier, they form a cementum-PDL-like structure (Seo et al., 2004). A specific STRO-1$^+$ stem cell population was isolated from PDL using FACS (Gay et al., 2007). This population of cells was able to form osteoblast, chondrocyte, and adipocyte-like cells *in vitro*. The varying potentials of the different populations of stem cells found in PDL promise that it will be a valuable cell source for periodontal tissue regeneration.

1.4.4   SALIVARY GLAND STEM CELLS

Loss of salivary gland function is a relatively common affliction. This loss is frequently attributed to autoimmune disease or to radiation therapy treatment. Salivary glands are composed of several different cell types with distinct functions, such as acinar cells, myoepithelial cells, and ductal cells. This tissue also contains a population of stem cells that are involved in the maintenance and repair of damaged tissue. Salivary gland stem cells have been isolated and characterized by their expression of c-kit, Sca-1, and Musashi-1, three previously identified cell surface markers known to be expressed on stem cells. These isolated cells were able to form acinar-like cells *in vitro*, and *in vivo* studies showed that when they were transplanted back into damaged salivary glands, they were able to restore tissue function (Lombaert et al., 2008).

1.4.5   BONE MARROW STROMAL CELLS

Bone marrow stromal cells (BMSC) are the adherent cells from bone marrow and have been shown to contain a subpopulation of mesenchymal stem cells. At this time, this population of cells is the most widely studied since they are easily isolated from autologous sources and have a large potential for expansion. BMSC have been shown to differentiate *in vitro* down several different lineages, including bone, cartilage, fat, tendon, and muscle (Pittenger et al., 1999; Alhadlaq and Mao, 2004). Currently, the differentiation of BMSC to a specific cell type *in vitro* can be managed by controlling the culture environment for growth factors and mechanical stimuli. Studies in which these cells have been subcutaneously implanted into immunocompromised mice with a hydroxyapatite carrier have shown that BMSC can form bone and bone marrow complexes. Although BMSC have relatively low proliferation potential and limited capacity to form dental specific tissue such as dentin or cementum (Scheller et al., 2009), their capacity to form bone-like and ligament-like tissues make them an important tool for the repair of both alveolar bone defects and ailments associated with the periodontal ligament.

BIBLIOGRAPHY

Alhadlaq, A. and J. J. Mao (2003). "Tissue-engineered neogenesis of human-shaped mandibular condyle from rat mesenchymal stem cells." *Journal of Dental Research* **82**(12): 951–956. DOI: 10.1177/154405910308201203 5

Alhadlaq, A. and J. J. Mao (2004). "Mesenchymal stem cells: Isolation and therapeutics." *Stem Cells and Development* **13**(4): 436–448. DOI: 10.1089/scd.2004.13.436 12

Alsberg, E., E. E. Hill, et al. (2001). "Craniofacial tissue engineering." *Crit Rev Oral Biol Med* **12**(1): 64–75. DOI: 10.1177/10454411010120010501 5

Athanasiou, K. A., G. G. Niederauer, et al. (1996). "Sterilization, toxicity, biocompatibility and clinical applications of polylactic acid/polyglycolic acid copolymers." *Biomaterials* **17**(2): 93–102. DOI: 10.1016/0142-9612(96)85754-1 2

Baum, B. J. and D. J. Mooney (2000). "The impact of tissue engineering on dentistry." *J Am Dent Assoc* **131**(3): 309–18. 6

Boussif, O., F. Lezoualc'h, et al. (1995). "A versatile vector for gene and oligonucleotide transfer into cells in culture and in vivo: polyethylenimine." *Proc Natl Acad Sci U S A* **92**(16): 7297–301. DOI: 10.1073/pnas.92.16.7297 8

Burdick, J. A., C. Chung, et al. (2005). "Controlled degradation and mechanical behavior of photopolymerized hyaluronic acid networks." *Biomacromolecules* **6**(1): 386–91. DOI: 10.1021/bm049508a 2

Cai, D. Z., C. Zeng, et al. (2007). "Biodegradable chitosan scaffolds containing microspheres as carriers for controlled transforming growth factor-beta1 delivery for cartilage tissue engineering." *Chin Med J (Engl)* **120**(3): 197–203. 7

Cen, L., W. Liu, et al. (2008). "Collagen tissue engineering: development of novel biomaterials and applications." *Pediatr Res* **63**(5): 492–6. DOI: 10.1203/PDR.0b013e31816c5bc3 2

Chen, F. M., R. Chen, et al. (2009). "In vitro cellular responses to scaffolds containing two microencapulated growth factors." *Biomaterials* **30**(28): 5215–24. DOI: 10.1016/j.biomaterials.2009.06.009 6

Chen, H., E. P. Frankenburg, et al. (2003). "Combination of local and systemic parathyroid hormone enhances bone regeneration." *Clin Orthop Relat Res*(416): 291–302. DOI: 10.1097/01.blo.0000079443.64912.18 6, 9

Cho, S. H., S. H. Oh, et al. (2005). "Fabrication and characterization of porous alginate/polyvinyl alcohol hybrid scaffolds for 3D cell culture." *J Biomater Sci Polym Ed* **16**(8): 933–47. DOI: 10.1163/1568562054414658 2

Choi, Y. H., F. Liu, et al. (1998). "Polyethylene glycol-grafted poly-L-lysine as polymeric gene carrier." *J Control Release* **54**(1): 39–48. DOI: 10.1016/S0168-3659(97)00174-0 8

Cochran, D. L. and J. M. Wozney (1999). "Biological mediators for periodontal regeneration." *Periodontol 2000* **19**: 40–58. DOI: 10.1111/j.1600-0757.1999.tb00146.x 7

Damrongsri, D., S. Geva, et al. (2006). "Effects of Delta12-prostaglandin J2 on bone regeneration and growth factor expression in rats." *Clin Oral Implants Res* **17**(1): 48–57. DOI: 10.1111/j.1600-0501.2005.01181.x 9

Dang, J. M. and K. W. Leong (2006). "Natural polymers for gene delivery and tissue engineering." *Adv Drug Deliv Rev* **58**(4): 487–99. DOI: 10.1016/j.addr.2006.03.001 8

Day, R. M., V. Maquet, et al. (2005). "In vitro and in vivo analysis of macroporous biodegradable poly(D,L-lactide-co-glycolide) scaffolds containing bioactive glass." *J Biomed Mater Res A* **75**(4): 778–87. DOI: 10.1002/jbm.a.30433 2

DeFail, A. J., C. R. Chu, et al. (2006). "Controlled release of bioactive TGF-beta 1 from microspheres embedded within biodegradable hydrogels." *Biomaterials* **27**(8): 1579–85. DOI: 10.1016/j.biomaterials.2005.08.013 7

Einhorn, T. A., R. J. Majeska, et al. (2003). "A single percutaneous injection of recombinant human bone morphogenetic protein-2 accelerates fracture repair." *J Bone Joint Surg Am* **85-A**(8): 1425–35. 6

Gay, I. C., S. Chen, et al. (2007). "Isolation and characterization of multipotent human periodontal ligament stem cells." *Orthod Craniofac Res* **10**(3): 149–60. DOI: 10.1111/j.1601-6343.2007.00399.x 11

Gonzalez, H., S. J. Hwang, et al. (1999). "New class of polymers for the delivery of macromolecular therapeutics." *Bioconjug Chem* **10**(6): 1068–74. DOI: 10.1021/bc990072j 8

Gronthos, S., M. Mankani, et al. (2000). "Postnatal human dental pulp stem cells (DPSCs) in vitro and in vivo." *Proc Natl Acad Sci U S A* **97**(25): 13625–30. DOI: 10.1073/pnas.240309797 3, 10

Honda, M., S. Tsuchiya, et al. (2007). "The sequential seeding of epithelial and mesenchymal cells for tissue-engineered tooth regeneration." *Biomaterials* **28**(4): 680–9. DOI: 10.1016/j.biomaterials.2006.09.039 2, 3

Ikeda, E., R. Morita, et al. (2009). "Fully functional bioengineered tooth replacement as an organ replacement therapy." *Proc Natl Acad Sci U S A*. DOI: 10.1073/pnas.0902944106 2, 3

Ishimatsu, H., C. Kitamura, et al. (2009). "Formation of dentinal bridge on surface of regenerated dental pulp in dentin defects by controlled release of fibroblast growth factor-2 from gelatin hydrogels." *J Endod* **35**(6): 858–65. DOI: 10.1016/j.joen.2009.03.049 6

Kamitakahara, M., C. Ohtsuki, et al. (2008). "Review paper: behavior of ceramic biomaterials derived from tricalcium phosphate in physiological condition." *J Biomater Appl* **23**(3): 197–212. DOI: 10.1177/0885328208096798 2

Kato, T., H. Kawaguchi, et al. (1998). "Single local injection of recombinant fibroblast growth factor-2 stimulates healing of segmental bone defects in rabbits." *J Orthop Res* **16**(6): 654–9. DOI: 10.1002/jor.1100160605 6

Kawaguchi, H., C. C. Pilbeam, et al. (1995). "The role of prostaglandins in the regulation of bone metabolism." *Clin Orthop Relat Res*(313): 36–46.

Kim, I. Y., S. J. Seo, et al. (2008). "Chitosan and its derivatives for tissue engineering applications." *Biotechnol Adv* **26**(1): 1–21. DOI: 10.1016/j.biotechadv.2007.07.009 2

Kirkham, J., A. Firth, et al. (2007). "Self-assembling peptide scaffolds promote enamel remineralization." *J Dent Res* **86**(5): 426–30. DOI: 10.1177/154405910708600507 2, 3

Krebsbach, P. H., S. A. Kuznetsov, et al. (1997). "Bone formation in vivo: Comparison of osteogenesis by transplanted mouse and human marrow stromal fibroblasts." *Transplantation* **63**(8): 1059–1069. DOI: 10.1097/00007890-199704270-00003 5

Kubarev, O. L., V. S. Komlev, et al. (2007). "Bioactive composite ceramics in the hydroxyapatite-tricalcium phosphate system." *Doklady Chemistry* **413**: 72–74. DOI: 10.1134/S0012500807030044 4

Kuo, T., A. Huang, et al. (2008). "Regeneration of dentin-pulp complex with cementum and periodontal ligament formation using dental bud cells in gelatin-chondroitin-hyaluronan tricopolymer scaffold in swine." *J Biomed Mater Res A* **86**(4): 1062–8. 2, 3

Laino, G., R. d'Aquino, et al. (2005). "A new population of human adult dental pulp stem cells: a useful source of living autologous fibrous bone tissue (LAB)." *J Bone Miner Res* **20**(8): 1394–402. DOI: 10.1359/JBMR.050325 11

Lombaert, I. M., J. F. Brunsting, et al. (2008). "Rescue of salivary gland function after stem cell transplantation in irradiated glands." *PLoS One* **3**(4): e2063. DOI: 10.1371/journal.pone.0002063 12

Lu, L., M. J. Yaszemski, et al. (2001). "TGF-beta1 release from biodegradable polymer microparticles: its effects on marrow stromal osteoblast function." *J Bone Joint Surg Am* **83-A Suppl 1**(Pt 2): S82–91. 7

Ma, P. X. (2008). "Biomimetic materials for tissue engineering." *Adv Drug Deliv Rev* **60**(2): 184–98. DOI: 10.1016/j.addr.2007.08.041 2

Mankani, M., S. Kuznetsov, et al. (2008). "Creation of new bone by the percutaneous injection of human bone marrow stromal cell and HA/TCP suspensions." *Tissue Eng Part A* **14**(12): 1949–58. DOI: 10.1089/ten.tea.2007.0348 2, 4

Marks, S. C., Jr., and S. C. Miller (1994). "Local delivery of prostaglandin E1 induces periodontal regeneration in adult dogs." *J Periodontal Res* **29**(2): 103–8. DOI: 10.1111/j.1600-0765.1994.tb01097.x 9

Miller, S. C. and S. C. Marks, Jr. (1994). "Effects of prostaglandins on the skeleton." *Clin Plast Surg* **21**(3): 393–400. 9

Miura, M., S. Gronthos, et al. (2003). "SHED: Stem cells from human exfoliated deciduous teeth." *Proceedings of the National Academy of Sciences of the United States of America* **100**(10): 5807–5812. DOI: 10.1073/pnas.0937635100 10

Moioli, E. K., L. Hong, et al. (2006). "Sustained release of TGFbeta3 from PLGA microspheres and its effect on early osteogenic differentiation of human mesenchymal stem cells." *Tissue Eng* **12**(3): 537–46. DOI: 10.1089/ten.2006.12.537 7

Moore, K. A. and I. R. Lemischka (2006). "Stem cells and their niches." *Science* **311**(5769): 1880–1885. DOI: 10.1126/science.1110542 9

Nakashima, M. (1994). "Induction of dentine in amputated pulp of dogs by recombinant human bone morphogenetic proteins-2 and -4 with collagen matrix." *Arch Oral Biol* **39**(12): 1085–9. DOI: 10.1016/0003-9969(94)90062-0 6

Nakashima, M. and A. Akamine (2005). "The application of tissue engineering to regeneration of pulp and dentin in endodontics." *J Endod* **31**(10): 711–8. DOI: 10.1097/01.don.0000164138.49923.e5 7

Pittenger, M. F., A. M. Mackay, et al. (1999). "Multilineage potential of adult human mesenchymal stem cells." *Science* **284**(5411): 143–7. DOI: 10.1126/science.284.5411.143 12

Prescott, R. S., R. Alsanea, et al. (2008). "In vivo generation of dental pulp-like tissue by using dental pulp stem cells, a collagen scaffold, and dentin matrix protein 1 after subcutaneous transplantation in mice." *J Endod* **34**(4): 421–6. DOI: 10.1016/j.joen.2008.02.005 6

Rohanizadeh, R., M. V. Swain, et al. (2008). "Gelatin sponges (Gelfoam) as a scaffold for osteoblasts." *J Mater Sci Mater Med* **19**(3): 1173–82. DOI: 10.1007/s10856-007-3154-y 2

Rutherford, R. B., M. E. Ryan, et al. (1993). "Platelet-derived growth factor and dexamethasone combined with a collagen matrix induce regeneration of the periodontium in monkeys." *J Clin Periodontol* **20**(7): 537–44. DOI: 10.1111/j.1600-051X.1993.tb00403.x 6

Schek, R. M., J. M. Taboas, et al. (2005). "Tissue engineering osteochondral implants for temporo-mandibular joint repair." *Orthod Craniofac Res* **8**(4): 313–9. DOI: 10.1111/j.1601-6343.2005.00354.x 5

Scheller, E. L. and P. H. Krebsbach (2009). "Gene therapy: design and prospects for craniofacial regeneration." *J Dent Res* **88**(7): 585–96. DOI: 10.1177/0022034509337480 8

Scheller, E. L., P. H. Krebsbach, et al. (2009). "Tissue engineering: state of the art in oral rehabilitation." *Journal of Oral Rehabilitation* **36**(5): 368–389. DOI: 10.1111/j.1365-2842.2009.01939.x 5, 8, 9, 12

Seo, B. M., M. Miura, et al. (2004). "Investigation of multipotent postnatal stem cells from human periodontal ligament." *Lancet* **364**(9429): 149–155. DOI: 10.1016/S0140-6736(04)16627-0 11

Shi, S. and S. Gronthos (2003). "Perivascular niche of postnatal mesenchymal stem cells in human bone marrow and dental pulp." *J Bone Miner Res* **18**(4): 696–704. DOI: 10.1359/jbmr.2003.18.4.696 11

Takafuji, H., T. Suzuki, et al. (2007). "Regeneration of articular cartilage defects in the temporo-mandibular joint of rabbits by fibroblast growth factor-2: a pilot study." *Int J Oral Maxillofac Surg* **36**(10): 934–7. DOI: 10.1016/j.ijom.2007.06.007 6

Takemoto, S., N. Morimoto, et al. (2008). "Preparation of collagen/gelatin sponge scaffold for sustained release of bFGF." *Tissue Eng Part A* **14**(10): 1629–38. DOI: 10.1089/ten.tea.2007.0215 7

Tenenbaum, L., E. Lehtonen, et al. (2003). "Evaluation of risks related to the use of adeno-associated virus-based vectors." *Curr Gene Ther* **3**(6): 545–65. DOI: 10.2174/1566523034578131 8

Trojani, C., F. Boukhechba, et al. (2006). "Ectopic bone formation using an injectable biphasic calcium phosphate/Si-HPMC hydrogel composite loaded with undifferentiated bone marrow stromal cells." *Biomaterials* **27**(17): 3256–3264. DOI: 10.1016/j.biomaterials.2006.01.057 2

Tschon, M., M. Fini, et al. (2009). "In vivo preclinical efficacy of a PDLLA/PGA porous copolymer for dental application." *J Biomed Mater Res B Appl Biomater* **88**(2): 349–57. 2

Ueki, K., D. Takazakura, et al. (2003). "The use of polylactic acid/polyglycolic acid copolymer and gelatin sponge complex containing human recombinant bone morphogenetic protein-2 following condylectomy in rabbits." *J Craniomaxillofac Surg* **31**(2): 107–14. 6

Ulijn, R. V. and A. M. Smith (2008). "Designing peptide based nanomaterials." *Chem Soc Rev* **37**(4): 664–75. DOI: 10.1039/b609047h 2

Verma, I. M. and N. Somia (1997). "Gene therapy – promises, problems and prospects." *Nature* **389**(6648): 239–42. DOI: 10.1038/38410 8

Watt, F. M. and B. L. M. Hogan (2000). "Out of Eden: Stem cells and their niches." *Science* **287**(5457): 1427–1430. DOI: 10.1126/science.287.5457.1427 9

Wei, G., Q. Jin, et al. (2006). "Nano-fibrous scaffold for controlled delivery of recombinant human PDGF-BB." *J Control Release* **112**(1): 103–10. DOI: 10.1016/j.jconrel.2006.01.011 7

Witte, F., H. Ulrich, et al. (2007). "Biodegradable magnesium scaffolds: Part II: peri-implant bone remodeling." *J Biomed Mater Res A* **81**(3): 757–65. 2, 4

Xu, W., W. Zhang, et al. (2008). "Accurately shaped tooth bud cell-derived mineralized tissue formation on silk scaffolds." *Tissue Eng Part A* **14**(4): 549–57. DOI: 10.1089/tea.2007.0227 2, 3

Young, C. S., S. Terada, et al. (2002). "Tissue engineering of complex tooth structures on biodegradable polymer scaffolds." *J Dent Res* **81**(10): 695–700. DOI: 10.1177/154405910208101008 2, 3

CHAPTER 2

Tissue Engineering Alveolar Bone

Mona K. Marei, BDS, MScD, PhD

Mohamad Nageeb, BDS, MSc

Rania M. Elbackly, BDS, MSc

Manal M. Saad, BDS, PhD

Ahmad Rashad, BDS, MSc

Samer H. Zaky, BDS, PhD

2.1 CHAPTER SUMMARY

Alveolar bone is a unique tissue representing the most viable part of the tooth-supporting apparatus. It exists solely when dentition exists, and hence it derives its uniqueness and value from this synergistic coupling. Regenerating the supporting tooth apparatus continues to gain more focus because of increasing demands on placing implants in healthy bone providing enhanced osseointegration quality with implants that can function under load earlier and can last longer.

Regenerative dentistry offers the distinct advantage of approaching regeneration rather than repair using various strategies individually or combined. Cells, growth factors and scaffolds are crucial elements in any tissue regeneration process; their use has also extended to attempts to regenerate the alveolar bone.

The recent shift in thinking is to implement in vitro processes that mimic *in vivo* tissue development that focus on direct relationships between growth and differentiation processes in embryonic development and postnatal tissue regeneration, they also have foot print infuses in alveolar bone engineering.

Understanding the pattern and signals behind alveolar bone development and healing can provide insight for designing novel regeneration strategies with increasing precision.

2.2   TISSUE ENGINEERING ALVEOLAR BONE

2.2.1   INTRODUCTION

It is always the matter of role and function that is usually targeted through the preceded development, lifetime adaptation/remodeling, or even atrophy of any tissue or organ as an integrated part of our body system. It is also worthy to know that the bone biochemical/biomechanical environment either in embryogenesis or an adult is a complex system with various scenarios and interactions tooled with versatile cocktails of signaling and non signaling molecules along with their antagonists and regulators, thus to control angiogenesis, innervations, different cell functions (migration, recruitment, proliferation, differentiation and matrix production) and further ossification, remodeling and repair (Allori et al., 2008a,b,c, Parts I, II, III).

From a material science of view, bone is a calcified composite matrix that has a specific biochemical/biomolecular composition (organic/inorganic components) and is developed via a specific method of processing during growth (intramembraneous/endochondral). Bone types have the same structure, but they have different architectures (compact/cancellous) that determine their principle functions (load-bearing, protection of vital organs, nutritional, production of marrow cells, ion exchange, etc.), yet to give this material its bio-color, the cellular components included in bone (osteoblasts, osteocytes, osteoclasts, marrow cells) aided the unique living soul, observed as growth, repair, lifetime remodeling, and adaptation.

The alveolar process consists of an external plate of cortical bone, the inner socket of thick, compact bone and cancellous trabeculae interposed. Alveolar bone is intramembranous in origin (Zhang et al., 2003; Chen and Jin, 2010) and undergoes continuous remodeling by osteoblast and osteoclast activity. For any periodontal tissue engineering strategy, regeneration of alveolar bone is mandatory. Since it is the alveolar bone in which attachment of periodontal ligament fibers into the cementum takes place, coordination between both soft and hard tissue healing is required (Chen and Jin, 2010). Interactive mechanisms controlling the induction and healing of the alveolar bone are not well studied (Zhang et al., 2003) and in the existence of disease, become more complex.

Thus, to study throughout the living classes and species, there are different properties of specialized organs that are adapted to function. Talking about the alveolar bone *per se* or as a part of the functional peridontium will lead us to a minuet comparison based on the functional anatomy of human teeth versus other animal classes that primarily use their teeth to snag prey but do not share mastication (cyclic functional loads) as a primary function of their teeth.

In many lower vertebrates (teethed vertebrates), the individual teeth show several replacements throughout the animal's life (polyphyodonts): old tooth loss and new tooth growth (http://qanda.encyclopedia.com/question). Hence, considering this dental regenerative prospect could be helpful to researchers. Snakes, lizards, crocodilians, and tuataras all have teeth. However, their teeth vary in form, attachment, and whether they are shed or not. Teeth may be attached on the alveolar surface of the jaw (acrodont), on the inner side of the jawbone (pleurodont), or in sockets (thecodont) with/without ligamentous attachment (Debra et al., 1979; Kardong, 2006) (Table 2.1) (Fig. 2.1 a–d).

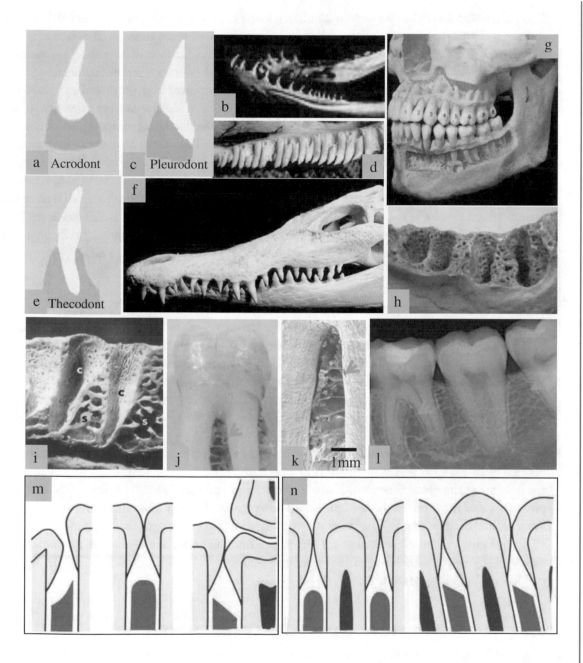

Figure 2.1: *Caption on next page.*

Figure 2.1: *Caption for figure on previous page.* Macroscopic and radiographic characteristics of the alveolar process. (a, b) diagram for acrodont teeth and macroscopic view of snake teeth. (c, d) diagram for pleurodont teeth and macroscopic view of lizard teeth. (e & f) diagram for thecodont teeth and macroscopic view of nile crocodile skull teeth. (g, h & i) macroscopic views of alveolar bone process in human, (g) is sagital view of human skull showing the alveolar process occupying roots of teeth (McMinn and Hutchings, 1991), (h) is an occlusal view for the alveolar process (the cribriform plate) after teeth extraction, (i) longitudinal section through the empty extracted sockets showing the thin lining compact bone of extracted socket on the upper part and the trabeculation of the cancellous bone as we go down to the roots portion of teeth "section L.S. in the premolar region of mandibular teeth, (Courtesy of Dr. Max Listgarten, http://www.dental.pitt.edu/informatics/periohistology/en/gu0403m.htm) (j, k, l) showing the horizontal trabecular pattern that appears radiographically white radiopaque lines separated by marrow spaces that looks radiolucent in radiographs. (molar region in human mandibular teeth) (m,n) the relationship of alveolar bone to tooth anatomy and position. (m) state of eruption, (n) inclination of teeth may all influence the contour of alveolar process (from Ritchey and Orban, 1953).

Mammals possess more developed peridontium, as they share cyclic functional loads, with only two dentition sets of teeth (deciduous and permanent; diphyodonts). Rabbits, for example, have periodontal ligament attached to the alveolar bone from only three sides, with the labial root of their lower incisors covered with enamel and only loose connective tissue filling the space to the alveolar bone. This is considered as a property that permits the *continuous eruption* in rodents due to the continuous incisal enamel wear, while others may have cementum pearls covering enamel; thus animal models for regenerating periodontium still possess a vital issue to be justified (Merzel et al., 2000; Moriyama et al., 2006).

In humans, periodontium is mainly mentioned as an integrated scene starting with the periodontal ligament (that will be discussed later in a separate chapter) due to its unique properties that fits in bearing the cyclic functional loads of mastication and chewing; this is why authors discuss properties and functions of the hard periodontal structures -alveolar bone and cementum- in most cases limited to attaching PDL. Indeed, the whole periodontal structure (alveolar bone, cementum, PDL and gingiva) are all important for intact, healthy functional dental supporting tissues *Periodontium*. However, alveolar bone may exist and function in human or animals with or without periodontal ligament attachment.

2.2.2   MACROSCOPIC FEATURES OF ALVEOLAR BONE

The alveolar process is that portion of jaw bones which supports and maintains the teeth. Although, no distinct anatomic separation can be distinguished between the alveolar process and the basal bone of the maxilla and mandible, the alveolar process *per se* does not develop until eruption of the teeth, when teeth fail to develop e.g., in anodontia, the alveolar process fails to form.

Table 2.1: Tooth/alveolus interfacial attachment in vertebrates.		
Acrodont (acro=end)	Sharks, snakes and tuataras	Teeth reside on the occlusal surface of the jaw bones in a very shallow socket (ankylosed at its base to the crest of the socket) with replacement teeth arise adjacent to the active teeth. The teeth are not firmly rooted and are easily lost and replaced.
Pleurodont (Pleur=side)	Most lizards	Teeth attached to the inner side of the jaw.
Thecodont (theca=cup)	Few reptiles (e.g., crocodiles)	Teeth with well developed ankylosed roots. Teeth replaced within the same socket. (The teeth are placed in a socket on the top of the jaw). A tooth may have a single root.
Thecodont (theca=cup)	Mammals and human	Teeth with well developed periodontal attachment (attaching roots to alveolar sockets), several roots as in the molars.

The alveolar process is divided anatomically into two parts; the alveolar bone proper, which is thin, modified compact bone forming the socket lining around the roots of teeth; the remainder of the alveolar process is referred to as supporting alveolar bone. This portion is made up of compact bone layer forming the vestibular and oral (*buccal and lingual cortical*) plates of the process, and spongiosa or cancellous bone that occupies the remaining space between the cortical plates and alveolar bone proper (Williams and Zager, 1978).

The alveolar bone proper is perforated by openings through which blood vessels, lymphatics and nerves pass to and from the periodontal ligament. This perforated part is referred to as cribriform plate, which is directly facing the tooth (Fig. 2.1 e,f,g,h,i).

Radiographically, the alveolar bone proper appears as a thin radiopaque line around the roots of the teeth known as lamina dura. Alveolar bone spongiosa appears as a series of thin, trabeculated white lines running horizontally (ladder like pattern) in mandibular posterior area, while differently oriented in maxillary area. These trabecular sheets are enclosing dark radiolucent spaces (marrow cavities) known as red hematopoietic marrow, spaced especially in maxillary and mandibular molar areas (Bhaskar, 1991) (Fig. 2.1 j,k,i).

In general, the anatomy of the teeth and their position in the skull determine the anatomy of the alveolar bone. The margin of the alveolar process coronally and interdental septa is rounded and may mimic the contour of the cementoenamel junction of the tooth. The width of the interdental septa is determined by tooth form present. Flat proximal tooth surfaces call for narrow septa, whereas in extremely convex tooth surfaces, wide interdental septa with flat crests are found (Ritchey and Orban, 1953).

An inclined or prominent tooth usually has the crest of the alveolar bone more apical than normal on the side of inclination with opposing side more coronal (Fig 2.1 m-n).

Blood supply to the alveolar bone is via branches of the alveolar artery. Blood vessels coursing over the surface of the oral and vestibular cortical plates as well as vessels in the interdental septum supply the alveolar bone, along with the gingival and PDL (Bhaskar, 1991).

2.2.3   MICROSCOPIC FEATURES OF ALVEOLAR BONE

The alveolar bone is in a constant state of remodeling; bone formation and resorption are occurring simultaneously and regulated by local and systemic influences. This physiological remodeling state allows the normal migration of teeth in a mesial direction when teeth proximal surfaces wear down (Nanci et al., 2003).

In alveolar bone, adjacent lamellae can be identified by the presence of the so-called cementing lines. These lines reflect where apposition or resorptive phases have occurred previously. If resorption was followed by opposition, the irregular scalloped line known as reversal line appears indicating the resorptive phase and demarcates between old and newly formed bones (McKee and Kaartinen, 2002) (Fig. 2.2).

Bone formation is indicated by the presence of cuboidal osteoblasts adjacent to a layer of osteoid, new bone which is undergoing mineralization. As bone formation proceeds, some of the osteoblasts become entrapped in the new formed bone and persist as osteocytes in lacunae.

When osteoclasts are actively participating in resorption, they could be seen in bony depressions known as hawship's lacunae on bone surface. At the ultra structural level osteoclasts are large multinucleated cells, while resting osteoblats appear as flat cells along the bone surface (Ritchey and Orban, 1953)).

The position of alveolar bone may change as a result of bone apposition and resorption (drift). The maxilla moves inferiorly in the skull due to bone resorption on its superior surface and bone deposition on its inferior surface, while mandibular bone resorps on the ramus and condyle as well.

2.2.4   MINERALIZATION OF ALVEOLAR BONE

Alveolar bone is a highly specialized mesodermal tissue, consisting of organic matrix and inorganic minerals component. The matrix is composed of a network of osteocytes and intercellular substance.

Theories of bone mineralization were first enunciated 80 years ago, while much more now is known about the process. In general, mineralization was first described in the hypertrophic chondrocytes of the growth plate as small vesicles with average diameter of 1000Å and were shown to start calcification in the endochondral bone formation (Bonucci, 1967).

Nucleating sites very similar to vesicles have been described in intermembranous fetal bone formation. These extrusions from osteoblasts into the preosseous matrix formed small globules on

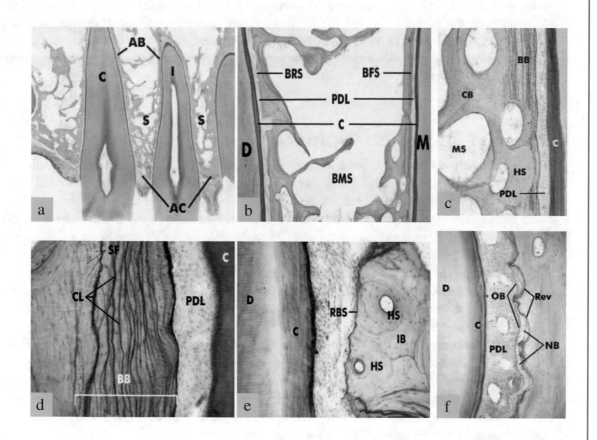

Figure 2.2: Microscopic view of alveolar bone remodeling during physiological mesial drifting of teeth (a & b) mesio-distal histological section through maxillary lateral incisor and cuspid teeth, Mesial is to the right of the micrograph. AB, alveolar bone, AC, alveolar crests, C, canine, I, lateral incisor, S, interdental septum, BFS abone-forming surface, BRS abone-resorbing surface, (b) enlarged view of (a), BMS abone marrow spaces, C acementum, D adistal side, M amesial side, PDL aperiodontal ligament. (c-f) Represent the alveolar bone surface of interdental septa between two neighboring teeth: (d) Bundle bone on the mesial surface of an interdental septum (BB) bundle bone, CL cement line, SF sharply fibers, PDL: periodontal ligament. (e) Bone surface during a resting phase. (f) Predominantly bone-resorbing surface during a phase of bone apposition RBS: predominantly bone- resorpting surface during a phase of bone apposition. NB : layer of newly formed bone, OB: osteoblasts, HS: haversian system, REV: reversal line.

Histological illustration was taken with permission from Courtesy of Dr. Max Listgarten. http://www. dental.pitt.edu/informatics/periohistology/en/gu0403m.htm which nucleation of crystallites bone nodules next formed. It was shown that calcium and phosphate accumulated in the mitochondria of bone forming cells are secreted in micropackets of amorphous calcium phosphate then go to calcification part where they dissolve and form apatite crystals on the matrix vesicles. These crystalline clusters grow and firmly attached to organic matrix presumably collagen fibers (Bernard and Pease, 1969).

2.2.5   ALVEOLAR BONE FUNCTIONS

Alveolar bone shares any other bone in some general properties, it, consequently, has common functions. Alveolar bone develops initially as a protection for the soft developing teeth and latter (as the roots develop) as a support to the teeth through Gomphosis (the unique fibrous joint that binds teeth to alveolar bone proper) and allows a minimal functional motion. Finally, as the teeth are lost, the alveolar bone resorbs (Avery and Chiego, 2006). Young alveolar bone is dense bone with smooth walled sockets, while aged alveolar bone is osteoporotic with a rough, jagged socket wall, fewer viable cells in lacunae, marrow tissue infiltrated by fat cells, and thus diminished tooth support (Avery and Chiego, 2006). In addition, it is related to the body systemic as well as local conditions. Some studies confirmed a direct proportion of skeletal osteoporosis to decrease bone mineral density BMD of interdental and basal bone (Marei et al., 2002, 2003b).

Tooth function is a prerequisite for the maintenance of the alveolar bone and cementum. Loss of alveolar bone (metabolic disease/osteoporosis/aging/ periods of inactivity) is always accompanied by loss of periodontal fibers. Loss of PDL, due to injury (e.g., tooth avulsion and replantation), could cause ankylosis, where alveolar bone becomes directly bonded to tooth that usually causes infraocclusion for the specific ankylosed tooth or even impairs a segmental vertical face growth (Saffar et al., 2000; Malmgren and Malmgren, 2002; Kawanami et al., 1999).

Alveolar crest always maintains 1.8 to 2 mm gingival area called "biological width" from alveolar crest to cervical line of the tooth in which free alveologingival fibers are inserting in the marginal alveolar crest and are radiating coronary into the gingival lamina propria (Marks and Schroeder, 1996). This area must be maintained biologically and mechanically free, thus any longstanding biological or mechanical impairment is kept out of this area through the withdrawal of this system to more apical position to preserve this defensive part through re-establishment of its fibers with subsequent vertical alveolar crest resorption.

From a biomechanical point of view, bone shape is fashioned into three-dimensional geometric and architectural masterpieces of bioengineered minimal mass, optimized in size and shape according to whether the main function is as a lever or as a spring (Dalle et al., 2005). For load bearing and leverage, the need for stiffness is favored over flexibility by the fashioning of mineralized tissue into long bones with a marrow cavity displacing the mineralized cortex distant from neutral bone axis. Vertebral bodies, spring-like shock absorbers on which stiffness is compromised for flexibility, showing an open-celled porous cancellous structure able to deform and return to its original size and shape without cracking (Seeman, 2003). It is logic to consider the alveolar bone of the spring type thus act as a force cushion conveying the masticatory forces (cyclic functional loads) to the basal bone through force trajectories to the basal bone (Gerstner and Cianfarani, 1998).

Alveolar bone has versatile functions, some of them are generalized for bone structures, including mechanical (protection, shape) and metabolic (minerals/growth factors, storage, and ionic exchange). Other functions are restricted to site with its specialized role (summarized in Table 2.2).

From clinical aspect, alveolar bone serves the following functions:

- Radiographic and clinical alveolar crest level determines the degree and fate of periodontitis.

- Lamina dura is an important diagnostic landmark in determining the health of the bony attachment of PDL fibers and socket lining wall support. Loss of density usually means infection, trauma, inflammation, and resorption of this bony socket lining (Avery and Chiego, 2006), although others consider lamina dura as superimposition of the image (Berkovitz et al., 2002).

- Decreased bone density (osteopenia) of alveolar process is noted when there is tooth inactivity with an absent antagonist.

- As oral surgeons have recognized maxillary tuberosity as a precious mine for autogenous bone graft for alveolar bone augmentation around dental implants, orthodontists also have been using maxillary tuberosity in anchoring of mini implants and distallization of maxillary molars.

- Extraction of an ankylosed tooth usually leads to loss of attached bone; the thin buccal plate of the maxilla and socket healing become particularly jeopardized, i.e., with defective horizontal and vertical dimensions (Malmgren et al., 2006).

- Porosity and permeability of maxillary alveolar bone permits local anesthesia infiltration to maxillary teeth. On the other side, mandibular or mental nerve block techniques are the only way to access mandibular teeth anesthesia due to thick, dense surrounding bone plates.

2.3   THE PROBLEM OF ALVEOLAR BONE RESORPTION

Alveolar bone resorption is a serious and common problem, especially, in edentulous patients, where alveolar ridge atrophy complicates the required efficient support and functioning of any prosthetic appliance. Since the alveolar bone development and maintenance are basically related to the existence of dentition to which is anchored, proper functioning and intimate interaction could be demonstrated and reflected in bone functional plasticity in response to all forms of structural and physiological changes associated with teeth. Such bone remodeling activity starts as early as tooth development; eruption, position changes accompanied with facial growth, and minor movements in adaptation to variable mechanical forces continue throughout life (Avery, 2001).

Property	Function
Table 2.2: Functions of alveolar bone related to its properties.	
Calcified bone matrix	-Protection, support to teeth, shape of the arch (mechanical). -Ion exchange and calcium storage (metabolic). -Growth factors storage (metabolic).
Periodontal attachment	-Periodontal ligament attachment (sharpy's fibers) to bundle bone. -Gingival attachment of a tooth and alveolar bone (attached gingival).
Continuous remodeling	-Compensates for root growth and functional toot wear. -Changes in positional relations of primary and permanent teeth. -Facial growth & repositioning of teeth. -Permits orthodontic movement; bone resorption (compression side), bone deposition and remodeling (tension) side; (easily resorped than cementum).
Muscle attachment	-Labial and buccal muscles (Buccinators muscle attachment).
Maintain biological width	Defensive mechanism of alveolar crest against biological or mechanical impairment.
Sinus approximation	-Limiting over expansion of the floor of the maxillary sinus.
Maxillary tuberosity	-Source for autologous bone graft in oral surgery. -Fixation of mini implants (Anchorage) for maxillary molar distallization in orthodontic treatment. -Mandatory in prosthetic rehabilitation (complete dentures).
Alveolar process anatomy	-Used in class diagnosis through lateral cephalometric x-ray views before orthodontic treatment (point A).
Cellular contents	-Remodeling, tissue integrity preservation and regeneration (stem cells and osteoprogenitors, osteoblasts, osteoclasts and osteocytes). -Immunologic.

Following tooth extraction, the empty dental alveoli fill-up with blood clots and, sequentially, go through normal cascade of wound healing with its characteristic histodynamic features. However, alveolar bone in the empty socket undergoes resorptive phenomenon. Few days after tooth extraction, bone resorption begins to appear at the alveolar crest and interradicular regions of the socket (Rivera-Hidalgo, 2001). Two months later, when the socket has been filled with newly formed bone, it is mainly trabecular bone that has been developed on a collagenous matrix frame, firstly formed during socket wound healing, and starting from its base occlusally (Todo, 1968; Jahangiri et al., 1998). The healing socket's new vertical height and bone contour never reaches the original ones (Garg, 2004; Marei et al., 2005). This alveolar bone resorption continues with a slower rate throughout life and may go well below, where the apices of the teeth existed; it has been named years ago as the residual ridge resorption or RRR (Atwood, 1971).

Microscopic studies have revealed evidence of osteoclastic activity on the external surface of the crest of residual ridges; the scalloped margins of Howship's lacunae sometimes contain visible osteoclasts: bone resorptive cells. However, there have been no longitudinal studies that report a spontaneous increase in ridge size, new bone formation, or presence of reversal lines on the external surface of the residual ridge (Atwood, 1971).

The pattern and timing of alveolar ridge resorption, following tooth loss, have been documented and classified for totally edentulous maxilla and mandible, with certain amount of variability occurring during the first 6 months (Cawood and Harvell, 1988). These conditions have significant impact on the functional and esthetic outcome of treatment; therefore, bone maintenance after dental extraction should provide clinicians with a more suitable implant position, which is known as ridge preserving techniques (Ashman, 2000).

It has been reported that ridge preservation immediately after extraction prevents 40-60% of jaw bone atrophy that normally takes place 2-3 years post extraction (Schwartz-Arad et al., 2000). While alveolar ridge resorption could be age related physiological changes, systemic diseases that affect the bone structure and architecture would certainly demonstrate similar manifestations in the alveolar bone (Hildebolt, 1997) as a part of the whole skeleton. Other systemic factors, including hormonal disorders and metabolic problems with a consequent low bone density are directly correlated to higher rate alveolar ridge resorption (Klemetti, 1996; Knezovie-Zlatarie et al., 2002).

Local factors concerning bone quantity and quality, applied mechanical forces, particularly those related to prosthetic appliances wearing, are important aspects to be considered. It has been reported by several studies that denture wearers show variable rates of residual ridge resorption in maxillary and mandibular arches, due to surface area, force distribution, and prosthesis design related reasons (Wyatt, 1998; Lopez-Roldan et al., 2009).

Another regional influential factor in the alveolar bone status is the periodontal ligament (PDL) condition. It is well known that the periodontal tissue has an essential metabolic role in bone formation and maintenance, so PDL pathological changes with resulting endotoxins have a direct impact on ridge resorption (Hausmann et al., 1970). Furthermore, and from the biochemical point of view, prostaglandins synthesized by periodontal tissue have been suggested to be one of the most important key mediators linking the mechanical stimuli to the resorptive activity in the alveolar bone (Yamasaki et al., 1980).

2.4   ALVEOLAR BONE REGENERATION: CONVENTIONAL AND CURRENT THERAPY

Historically, there were many techniques and materials considered for grafting fresh extracted sockets to minimize the deformities in ridge contour later on. Sources of bone grafts, both autogenous and alloplastic, have been studied and compared. Fresh autogenous cancellous bone is ideal because it supplies living, immuno-compatible bony cells that integrate with the surrounding host bone and are essential for osteogenesis (Marks, 1993; Boyne, 1997).

Autogenous bone grafts have been used to treat patients with alveolar bone defects since the beginning of the 20th century. The usual sites for harvesting autogenous bone have included iliac crest, cranium, chin, and rib. Autogenous bone grafting due to superior patency for long term physiological function remains the gold standard technique. This invasive treatment is associated with high donor site morbidity in addition to being insufficient for large critical size defect (Arrington et al., 1996; Younger and Chapman, 1989). Allograft and xenograft implants can partly compensate these problems, but they are susceptible to rejection and infection (Betz, 2002; Horch and Pautke, 2006; Nevins and Mellonig, 1992).

It used to be believed that bone graft taken from membranous bone origin (e.g., cranial and mandibular) resorbs less over time than bone graft taken from endochondral bone origin (e.g., iliac crest). The current knowledge supports the fact that maintenance of volume is the result of bone micro architecture (Ozaki and Buchman, 1998; Ozaki et al., 1999). Grafting materials, for example, cancellous porous bovine bone mineral (PBBM) particles were shown to enhance new osseous tissue regeneration in extraction sockets (Artzi et al., 2000).

Ridge reduction was shown to have decreased in fresh alveolar sockets treated with nonresorbable hydroxyapatite (HA) (Nemcovsky and Serfaty, 1996), bioabsorbable membrane made of poly-L-lactic acid/polyglycolic acid (PLA/PGA) polymer (Lekovic et al., 1998), synthetic osseous graft (Murray, 1998), or calcium phosphates (Schneider, 1999) in advanced bone loss cases around periodontally involved teeth.

To avoid compromised results in implant fixture placement and to maximize function and aesthetics, a combination of xenograft and a cortical chin graft (Wiesen and Kitzis, 1998) or bone grafting and a biodegradable membrane (Yang et al., 2000) were used. A higher density of tissue was observed under an expanded polytetrafluoroethylene (e-PTFE) membrane, in comparison with the use of demineralized freeze-dried bone, or hydroxyapatite adjacent to immediate endosseous implants (De Vicente et al., 2000; Chiapasco et al., 1999; Fowler et al., 2000). Immediate post extraction implants were thought to have a high percentage of bone implant contact and to enhance esthetic results (Corelini et al., 2000; Wheeler et al., 2000; Vogel and Wheeler, 2001). It has also been suggested by several investigators that the use of bioactive glass particulate graft alone, or in combination with e-PTFE and calcium sulfate, or with implant placement is of some benefit in preserving alveolar ridge dimensions after tooth extraction (Sy, 2002; Camargo et al., 2002; Norton and Wilson, 2002). Synthetic copolymer has served as an osteoconductive material in bony defects around titanium implants (Forum and Orlowski, 2000).

Regeneration of skeletal tissues has been recognized as a new means for reconstruction of skeletal defects, e.g., osteogenic distraction. The technique was first introduced by Gavriel Ilizarov in 1952, while human maxillary distraction cases were first reported in 1992 (Ilizarov and Ledyaev, 1992; McCarthy et al., 1992).

The principle of distraction osteogenesis is based on the regeneration of new bone that develops when tension forces are applied, which is mainly intramembranous. After a low power osteotomy is performed, distraction osteogenesis begins with the formation of a hematoma between and around the bone segments, then bone necrosis occurs at the end of the fracture segments. An ingrowth of vasoactive elements and capillaries for the restoration of blood supply forms a soft callus.

Tension is then applied to the soft callus and a dynamic microenvironment is created. Pluripotential mesenchymal cells are activated into fibroblasts and osteoblasts, and type I collagen is laid down parallel to the vector of distraction. Bony trabeculae grow into the fibrous area from the periphery, parallel to the line of tension that occurs during the distraction phase. A bridge of immature bone thus forms across the distraction gap. A poorly mineralized, radiolucent fibrous interzone is located in the middle of the distraction gap, where the influence of tensional stress is maximal. The interzone functions as the center of fibroblast proliferation and fibrous tissue formation. During the consolidation phase, bone remodeling begins, and the newly regenerated tissue eventually ossifies, then undergoes maturation similar to the native bone (Tibesar et al., 2006; González-García et al., 2009).

Distraction osteogenesis decreases the need for bone grafting for large (>10 mm) mandibular advancements; one can achieve 20 mm or more of advancement without a bone graft and the associated donor site morbidity, scarring, and potential for infection. Greater patient acceptance exists with this procedure, especially with the development of low-profile intraoral devices (Kunz et al., 2005; Malkoç et al., 2006).

With technologic advancements, distraction devices have become smaller and more sophisticated than earlier versions. Distraction osteogenesis may even be teamed with endoscopic techniques to allow the placement of these devices with minimal surgery. Preliminary studies of rabbits have shown that distraction performed in the presence of recombinant human bone morphogenetic protein placed into the distraction site accelerates bone formation (Yonezawa et al., 2006).

With experience, the overall complication rate is low. Complications include the following: fibrous nonunion or premature union of bone, infection that may hinder osteogenesis, noncompliant patient with treatment failure, scarring of the skin with external devices, hardware failure, and malocclusion because of improper vectors. Although, distraction osteogenesis technology demonstrated increase on the dimensions of maxillary and mandibular bone that replace the historical ridge augmentation technique, yet it did not deal with the alveolar process preservation post extraction, but it involves the residual ridge after total alveolar process resorbs.

Within this context, the field of tissue engineering emerged twenty years ago, its principles derived from embryonic events and tissue remodeling because they provide the necessary bluepoint for the design of tissue engineering systems. In addition, tissue engineering has its applications in developmental biology by providing research models and advanced technologies to study and save biological species (Langer and Vacanti, 1993; Vunjak-Novakovic and Kaplan, 2006).

Tissue engineering was initially defined by the attendees of the first NSF sponsored meeting in 1988 as "application of the principles and methods of engineering and the life sciences toward the fundamental understanding of structure/function relationships in normal and pathological mammalian tissues and the development of biological subsrirutes to restore, maintain, or improve functions" (Skalak and Fox, 1988).

As the field advances toward functional tissue engineering, regenerative medicine, it improves our understanding of many aspects of normal tissue regeneration by providing high insight into models for controlled biological studies. Due to the fact that every tissue type has its own set of mechanical, hydrodynamic and electrical forces, chemical gradient, structure and morphology, and designs of tissue engineering systems should address general and tissue specific needs (Ingber et al., 2006).

2.5   ALVEOLAR EXTRACTION SOCKET: A MODEL FOR ALVEOLAR BONE WOUND HEALING

The regeneration of bone encompasses anatomical as well as functional restoration of tissue continuity, resulting in new bone formation with increase in the overall volume of the bone, thereby simulating the original bone in both quantity and quality. This is in contrast to repair in which only a restoration of the continuity of the injured tissues takes place (Al-Aql et al., 2008).

In the tooth extraction socket, the repair process displays a cascade of healing events reminiscent of intramembranous bone development and formation of bone in membrane protected defects of the alveolar ridge (Trombelli et al., 2008). Inflammation plays an important role in the healing cascade following extraction dictating the sequence of events that follow (Korpi et al., 2009), including inflammatory cells recruitment, that produce various proteinases and cytokines which govern the wound healing process. Cytokines such as MMP-8 (collagenase-2) have been found to regulate collagen metabolism, inflammatory cell recruitment, and cytokine profile in tooth extraction wound healing in mice (Mendes et al., 2008).

Within four weeks following extraction, there is already an organization of the fibrin blood clot, which is subsequently replaced by fibrous tissue and blood vessels (Amler, 1969; Lang et al., 2003; Mendes et al., 2008). Between 4 and 8 weeks post-extraction, there is a proliferation of osteogenic tissue, first appearing as woven bone, which then matures into trabecular bone, culminating in a remodeling process (Devlin and Sloan, 2002; Trombelli et al., 2008) (Fig. 2.3).

In spite of the rapid progressive replacement of granulation tissue by woven bone, the remodeling process of extraction sockets in humans appears to be quite slow. Substantial amounts of mineralized lamellar bone can be seen at the 6 month interval, yet, between six and twelve months there is a noticeable reduction in the amount of bone, and even at twenty-four weeks post-extraction, the bone organization and architecture is incomplete (Trombelli et al., 2008).

Although extraction socket healing has been extensively studied histologically and clinically, cellular and molecular mechanisms leading to bone formation and remodeling are not fully understood. Of those, the origin of the osteoblasts remains elusive, although in the rat extraction socket they are believed to have originated from the residual periodontal ligament fibroblasts migrating into the coagulum (Ramakrishnana et al., 1995).

In human beings immediately after injury, the healing cascade begins as early as ten days post-extraction, and it has been recently shown that Runx2, which is a marker of osteogenic differentiation, was found to be strongly expressed by osteoblasts in the marrow, by osteoprogenitors and a subpopulation of mature endosteal osteoblasts after 2 weeks in the extraction socket. This demonstrates that both the periodontal ligament and the bone marrow contribute to the osteoprogenitor populations responsible for bone formation (Devlin and Sloan, 2002). Cues governing the healing process in extraction sockets may provide insight to develop strategies for alveolar bone engineering (Kanyama et al., 2003).

2.6   TISSUE ENGINEERING ALVEOLAR BONE

The current knowledge dealing with regeneration of alveolar bone in adult involves three existing alternatives: an acellular approach, a cell-based approach, and a combined approach.

The cellular approach aims to facilitate the repair activity of native cell populations. They may consist of a nude matrix or of a matrix containing biological signaling molecules that will recruit and trigger the differentiation of local stem cell populations, resulting in the formation of a functionally competent type of repair tissue.

Cell-based therapy can involve adult stem cells, embryonic stem cells, or differentiated cells. If undifferentiated cells are used, they may either be directed to differentiate *in vivo* or be induced to do so *in vitro* prior to implantation. The cells can be applied either alone or in conjunction with a matrix. The combined approach is the most popular of the three alternatives. In this system, cell populations are trapped within a matrix that is functionalized with a biological signaling molecules. In all approaches, these systems may be introduced surgically or arthroscopically, as a solid entity or in an injectable form, under a great awareness of all ethical guidelines. We reviewed central paradigm for tissue engineering alveolar bone in three basic components:

1. Cellular key-players in alveolar bone engineering.

2. Growth factors for alveolar bone engineering: Intrinsic signals for bone development and repair.

3. The scaffold: the engineered guided tissue mimetic.

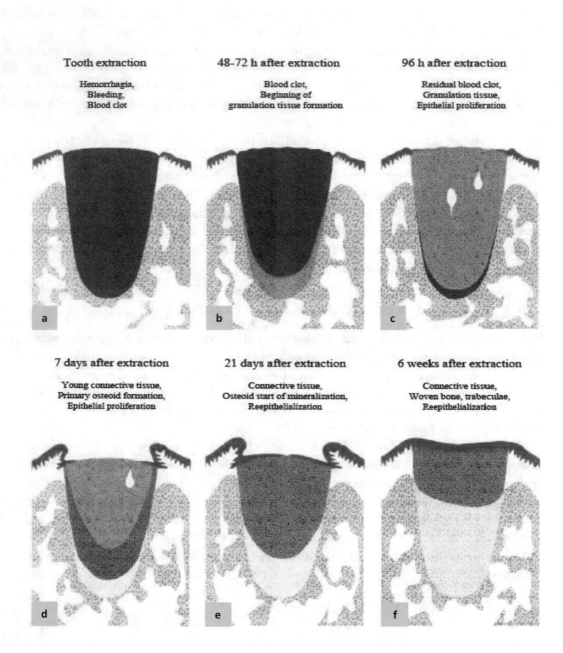

Figure 2.3: *Figure continues on next page.*

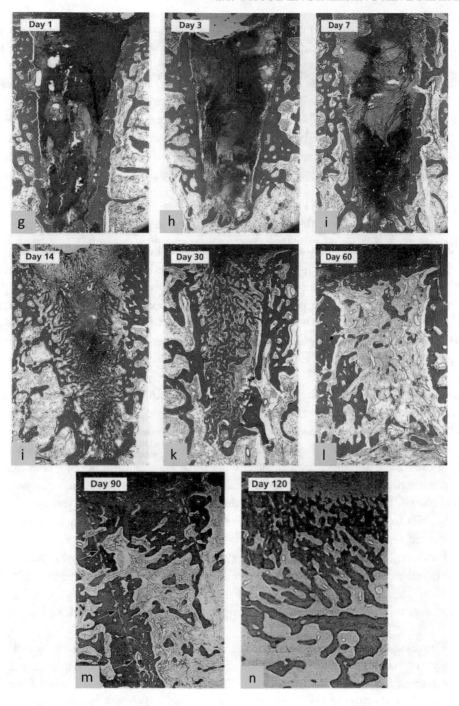

Figure 2.3: *Figure continues on next page.*

Figure 2.3: *Caption for figures on previous pages.* Cascade of alveolar bone wound healing of the extraction socket of teeth. (a-f) Illustration of events of wound healing in extraction socket at immediate post extraction and up to 6 weeks (adopted from Amler, 1969). (g-l): Microscopic view of the four phases of wound healing. (g) Empty socket is filled with blood clot. (h) Inflammatory cells migrated into the clot and process of wound cleansing initiated. (i) Vascular invasion and mesenchymal cells migration to the clot and granulation tissue formed. (j) Granulation tissue gradually replaced by a provisional connective tissue. (k & l) Formation of new woven bone. (m & n) woven bone modeled and remodeled into lamellar bone and marrow. (Adopted from Lang et al., 2003). Illustrations and figures are reprinted from Lindhe et al. (2008) Copyright, with permission from Blackwell Publishing Co.

2.6.1   CELLULAR KEY-PLAYERS IN ALVEOLAR BONE ENGINEERING

Applying the tissue engineering approach, many attempts have investigated alveolar bone regeneration either in periodontal defects or around dental implants in different animal models using several types of cells populations. Bone Marrow Stromal Cells (BMSC), a heterogeneous population of mesenchymal progenitors occupying the stroma of bone marrow, also termed Mesenchymal Stem Cells (MSC) (Owen and Friedenstein, 1988) or Skeletal Stem Cells (SSC) (Robey and Bianco, 2006) have been reported to regenerate periodontal alveolar defects (Li et al., 2009), to preserve the alveolar bone height inevitably reduced after dental extraction (Marei et al., 2005), or to regenerate bone around dental implants (Weng et al., 2006), or to regenerate bone for periodontal ligament attachment around dental implant (Marei et al., 2009). Periosteal cells (Mizuno et al., 2006) (undifferentiated osteoprogenitors that are mobilized from the bone envelope to contribute to normal bone growth, healing, and regeneration), adipose-derived stem cells (Tobita et al., 2008) (undifferentiated mesenchymal cells in adipose tissue), and osteoblasts (Gallego et al., 2010) (end-differentiated bone forming cells) have also been applied for alveolar bone regeneration. Among cells of the periodontium, periodontal ligament stem cells (PDLSC), believed to be a sub-population of clonogenic progenitor cells that can differentiate into either cementum-forming cells (cementoblasts) or bone-forming cells (osteoblasts) in vivo (Huang et al., 2009), were able to regenerate the periodontal apparatus recovering the heights of alveolar bone (Liu et al., 2008; Seo et al., 2004).

However, in the ambiguity of the definitions of the previously mentioned subsets of stem/progenitor cells, the nature of their cellular constituents, whether they are to be classified as "genuine" stem or as progenitors, is still elusive. Fig. 2.4 aims to simplify the general agreed concept and fate of a postnatal (adult) stem/progenitor/end-differentiated cell (Zaky and Cancedda, 2009).

Thinking about its continuous remodeling and its engineering exigency by cell based therapy, primarily to reverse bone loss in periodontal disease, one couldn't help but question about the original cellular players involved in regeneration of alveolar bone, including the role of the bone marrow harbored in its trabecular spaces. Bone and bone marrow have always been regarded as a single functional unit with common progenitors that give rise to both bone-forming cells and hematopoietic

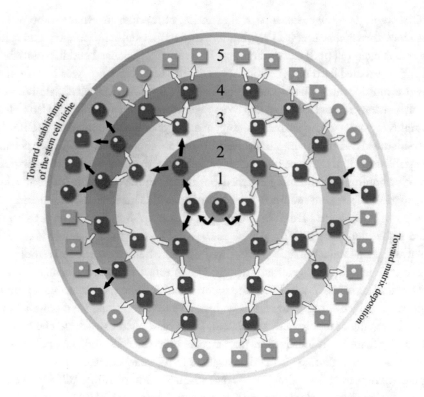

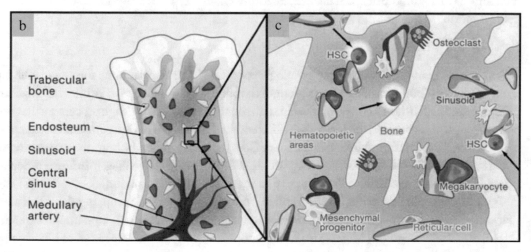

Figure 2.4: *Caption on next page.*

Figure 2.4: *Caption for figure on previous page.* Illustration of mechanisms that promote stem cells and niches maintenance throughout life: (a) Proliferation, differentiation, multipotency and self-renewal of an adult stem cell. A stem cell (red) can differentiate into several specialized enddifferentiated cells types (lineages - round or squared light gray, osteoblasts in case of bone, chondrocytes in case of cartilage), a property known as multipotency. The process occurs through asymmetrical cell division (black arrows) or symmetrical cell division (white arrows) passing by a progenitor genotype (dark gray, also called precursors or transient amplifying progeny), while maintaining, in each generation/population doubling (concentric circles) mother-identical cells that guarantee proper tissue turnover (self-renewal). These peculiar divisions results in the establishment of the stem cell niche (presented by the darker gradient background) from a side and the microenvironment promoting matrix deposition from the other side (presented by the lighter gradient background). Illustration modified from Zaky SH and Cancedda R. Engineering Craniofacial Structures: Facing the Challenge. J Dent Res. 2009. (b & c) A simplified illustration representing alveolar bone (b) and a trabecular space bearing a bone marrow niche (c). Bone marrow is a complex organ containing many different hematopoietic (HSC-blue) and non-hematopoietic cells (mesenchymal stem cells (MSC) and their daughter progenitors) (white). The interface of bone trabeculae and bone marrow is known as the endosteum. MSCs and HCSs are found mainly adjacent to sinusoidal blood vessels (red) and may reside in spatially distinct endosteal (osteoblastic) and perivascular (hematopoietic) niches that may or may not be locally or functionally equivalent. Osteoblasts and osteoclasts elaborate factors that regulate HSC maintenance and localization perivascular reticular cells and mesenchymal progenitors have also been proposed to elaborate factors that regulate HSC maintenance. Illustration reprinted with contextual legend modification from Cell, 22/132, Morrison SJ and Spradling AC, Stem cells and niches: mechanisms that promote stem cell maintenance throughout life, pages 598–611, Copyright (2008), with permission from Elsevier.

cells (Dominici et al., 2004). Under physiologic conditions, the alveolar bone turnover is the harmonious work of the bone-forming osteoblasts and the bone-eating osteoclasts. While the latter are end-differentiated cells originating from their circulating monocytes/macrophage precursors that home to bone through blood vessels, the former are differentiated from their periosteum/endosteum progenitors (Colnot, 2009) or from Mesenchymal Stem Cells (MSC) residing the bone marrow contiguous to blood sinusoids (Lee et al., 2007; Sacchetti et al., 2007) (specialized venules forming a reticular network of fenestrated vessels that allow cells to pass in and out of the circulation) (Garrett and Emerson, 2009). The concept of a specific, however still ill defined, environment, occupying the trabecular bone spaces and occupied by self-renewing, multipotent stem cells (Fig. 2.4 a), hematopoietic stem cells (HSC) and their daughters, Mesenchymal/hematopoietic progenitors associated to blood sinusoids, has been initially investigated three decades ago (Schofield, 1978) and recently studied under the name of Bone Marrow Niche (Fig. 2.4 b) (Morrison and Spradling, 2008).

THE NICHE CONCEPT

Evidently, this complex microenvironment, where different cell types support and impact on each others' survival, self renewal and differentiation, is a key-player in bone regeneration and its proper turnover (Garrett and Emerson, 2009; Charbord et al., 1996). Accordingly, to maintain this desirable microenvironment, it seems essential to transfer the progenitors-bearing niche with bone forming cells to the regeneration site or to guarantee the transplantation of a "genuine" stem cell-rich population that warrants regeneration of bone as well as its niche. The fact explains the rationale behind the autologous bone graft remaining the gold standard for bone regeneration (Rawashdeh and Telfah, 2008; Morselli et al., 2009). The autologous graft transfers "en block" trabecular bone and bone-depositing cells together with the niche-bearing trabecular spaces containing their own hierarchical meshwork of arterioles and capillaries as well as self renewing stem cells.

For the other option of delivering an enriched population of self-renewing cells, many markers have been proposed for selective enrichment of the stromal population by multipotent, self-renewing progenitors. Most convincing, multipotential MSC were defined as a specific population of perivascular cells with self-renewal capacities being positive to CD146 and negative to CD45 (non hematopoietic) cell surface markers. In a immunocompromised mouse model, this population was found to self-renew since they could generate multipotent CD146+ perivascular cells in serial transplantations, generating both bone and marrow. In contrast, cells in which CD146 was down-regulated, no self-renewal was observed. In the same model, bone, but not marrow, was formed in transplants of periosteal cells and human trabecular osteoblasts, evidencing the ability of a single CD146+ cell to establish both heterotopic bone and the hematopoietic marrow. In the implants, endothelial cells lining sinusoids were found to be of host origin (murine), while stromal cells at the albumin side of sinusoids, or interspersed among hematopoietic cells, were of donor origin (human) (Sacchetti et al., 2007).

The current view supports that this specific population of self-renewing cells is, in its big part, the pericytic population that reaches the regeneration site by invading the sinusoidal wall to which they are associated during early events of bone formation (Sacchetti et al., 2007; Deschaseaux et al., 2009). To date, although the niche residents (MSC, endothelial cells, osteoblasts, and HSC) are extensively described in vitro cultures, their in vivo native characterization are yet poorly understood (Deschaseaux et al., 2009).

Ultimately, studies on periodontal engineering (Nakajima et al., 2008) and on bone engineering in heterotopic model (Tortelli et al., 2010) have been reporting that upon transplantation of certain cells populations, tissue regeneration occurs by the recruited host cells in the implantation site and not by the implanted cells themselves. The particular allure of such reports lies in the ability of the implanted cells to alter the host response boosting the self-cells (host-own) to accomplish the regeneration task and continually maintain it (turnover) independently of the availability of the transplanted donor cells that would, sooner or later, undergo apoptosis after a certain period of "service".

Of an additional interest, some recent studies support the notion that bone regeneration by endochondral rather than intramembranous ossification would preponderate the re-establishment of the Stem Cell Niche (Sweeney et al., 2008; Chan et al., 2009). Apparently, the concept is, at present, too unripe to evade our understanding and to be investigated in the alveolar bone model whose regeneration and socket healing have been reported to be mainly through intramembranous or a mixture of intramembranous and endochondral ossifications (Nakajima et al., 2008; Ting et al., 1993) and were favored by intramembranous rather than endochondral grafts (Brugnami et al., 2009; Lu and Rabie, 2004). Interestingly, a study has reported "chondrogenic priming" of bone marrow stromal cells to boost bone repair by endochondral ossification should be more elaborated (Farrell et al., 2009).

It is reasonable to speculate that these recently investigated concepts crucial for engineering of tissues, i.e., the stem cell niches, host cells recruitment and the endochondral-associated niche regeneration, which are being studied in different experimental models, are also valid for the alveolar bone model. However, it has been observed, in the same individual, site-specific differences in the behavior of stem/progenitor cells, growth factors/cytokines and the extracellular matrix among different bones that have different embryological origins, different trabecular organization and marrow components and different (thus variable) adaptation to functional demands. As part of the neural crest-derived craniofacial complex under significant amount of masticatory stress and strain, the alveolar bone acquires its peculiarity from being under continuous periodontal pressure and tension, containing less red marrow compared to the axial load-bearing skeleton (Akintoye et al., 2006; Matsubara et al., 2005), hence, would arguably be engineered by adopting regeneration measures different from other cranial or axial bones (Zaky and Cancedda, 2009; Mao et al., 2006). Accordingly, alveolar bone-specific heuristic studies are of primary requisite to understand the particular molecular and cellular events that converse and act in conjunction for this intricate bone regeneration.

2.6.2   GROWTH FACTORS FOR ALVEOLAR BONE ENGINEERING: INTRINSIC SIGNALS FOR BONE DEVELOPMENT AND REPAIR

Growth factor profiles govern the pathways in which bone induction takes place. Bone induction can occur either directly from the mesenchyme or by first passing through a cartilage stage, which is then replaced by bone (Reddi, 2000). In the latter, first, mesenchymal cells undergo condensation, followed by differentiation of chondrocytes. Chondrocytes proliferate and produce extracellular matrix, then these proliferating chondrocytes in the central region of the cartilage undergo differentiation into hypertrophic chondrocytes, which synthesize a different cartilage extracellular matrix. The hypertrophic cartilage is then invaded by blood vessels, osteoblasts, osteoclasts and hematopoietic cells, resulting in formation of primary ossification centers. This is followed by hypertrophic cartilage matrix degradation and subsequent replacement by bone (Ripamonti et al., 2006). These processes have been shown to be mainly controlled by members of the transforming growth factor-β family (TGF-β).

(TGF-β) superfamily consists of a large number of growth and differentiation factors (Al-Aql et al., 2008). These include bone morphogenetic proteins (BMPs), transforming growth factor beta (TGF-β), growth differentiation factors (GDFs), activins, inhibins, and Müllerian inhibiting substances. Members of this family have been shown to have significant actions during intramembranous and endochondral ossifications pathways (Kaigler et al., 2006; Al-Aql et al., 2008). In particular, bone formation induced by BMPs recapitulates endochondral bone formation (Reddi, 2000).

Bone morphogenetic/osteogenic proteins (BMPs/OPs) control morphogenesis, organogenesis and epithelial/mesenchymal interactions. Studies of the mouse gene knockout system have revealed that BMP signaling is greatly involved in early embryonic patterning (Kishigami and Mishina, 2005).

During skeletal patterning and bone formation (bone modeling), complex processes involving more than one member of the BMP family take place. This is clearly demonstrated by the fact that genetically modified mice having inactivated BMP receptor type IB gene (BmprIB) are viable and exhibit defects that are largely restricted to the appendicular skeleton. This can be due to functional compensation from other BMPs residing in the same tissue, which highlights the importance of the synchrony between several growth factors (Tsumaki and Yhosikawa, 2005). This cell specific activation or inactivation of BMP signals in mice has shown that BMPs regulate differentiation of cells in both cartilage and bone (Axelrad and Einhorn, 2009), hence BMPs can be considered mesoderm inducers rather than bone inducers.

New bone arising from heterotopic implantation of lypholized bone segments is a demonstration of bone auto-induction by sequestered BMPs in the extracellular matrix of bone (Urist, 1965; Reddi, 2000; Ripamonti et al., 2006). For example, the endochondral ossification cascade elicited in defects of membranous calvarial bone in adult primates is evidence for bone development by induction, rather than merely by osteoconduction from the margins of bone defect (Ripamonti and Reddi, 1992). However, only a subset of BMPs, most notably BMP-2, -4,-6,-7, and -9 have osteoinductive activity; a property of inducing de novo bone formation by themselves (Kaigler et al., 2006; Al-Aql et al., 2008; Axelrad and Einhorn, 2009). BMPs 1, 3, and 12 appear to be incapable of these effects as they fail to induce alkaline phosphatase expression in osteoblastic cells (Axelrad and Einhorn, 2009).

Some studies have demonstrated that transgenic mice lacking BMP-4 and GDF-5 show reduced zones of proliferating chondrocytes. The involvement of BMPs in endochondral bone formation is clear, yet, the mechanisms dictating whether healing will occur via endochondral or intramembranous pathways are not (Tsumaki and Yhosikawa, 2005). In the early tooth germ, BMP-4 appears to be irreplaceable for alveolar bone formation and for initiation of tooth development. It has been hypothesized that BMP-4 may initiate osteogenic differentiation or act to regulate the differentiation of osteogenic cells in the alveolar bone (Zhang et al., 2003).

BMPs act directly on target cells to affect their survival, proliferation, and differentiation. When BMPs are secreted from cells, they can exert their actions locally, they may be bound up by extracellular antagonists present at the site of BMP secretion, or they may interact with extracellular matrix proteins that serve to sequester or enhance BMP activity by anchoring it to make it more available to target cells. Cell-specific production of BMPs or BMP receptors still remains unknown, yet several BMP antagonists have been recognized that are capable of binding to and blocking BMP receptors. When misexpression of these antagonists occurs, BMP signaling pathways are altered, thereby having effects on bone formation. Such BMP antagonists could also play a role, along with BMPs, in the regulation of bone matrix changes that coincide with the aging process. One such common molecule is BMP-3, which is an osteoblast product found in the bone matrix, and its deficiency has been found to induce increased bone formation in BMP-3 null mice (Rosen, 2006).

2.6.2.1 Alveolar Bone Healing and Fracture Repair: Similarities Between Two Models

In order to further explore the role of BMPs in bone formation, models of fracture healing and distraction osteogenesis have been studied (Kanyama et al., 2003). During bone fracture healing, bone morphogenetic proteins interact together as well as with other members of the transforming growth factor-beta family in an intricately synchronized fashion, eliciting specific responses that result in the formation of cartilage and bone. Cellular processes initiated include chemotaxis, mesenchymal cell proliferation and differentiation, angiogenesis, and production of extracellular matrix (Al-Aql et al., 2008; Reddi, 2000).

In fracture healing, several growth factors are involved which follow a time-course in which different signals are expressed and co-expressed at different times during the healing process (Dimitriou et al., 2005). In bone healing process, BMPs signaling appear to be the initiators, whereby BMP-2 mRNA shows maximal expression during the first 24 hours post-injury as BMP2 is necessary for post-natal bone repair and turnover. BMP-3, 4, 7, and 8 are mostly expressed during the phase of cartilage resorption and osteoblast recruitment for initiation of replacement bone formation. BMP-5 and 6 are expressed throughout the 21 days of the fracture healing process indicating their pivotal roles in both intramembranous and endochondral bone formation (Al-Aql et al., 2008).

BMP 2, 6, and 9 are the most effective inducers of mesenchymal cell differentiation into osteoblasts while remaining BMPs promote osteoblast maturation (Rauch et al., 2000; Al-Aql et al., 2008). Fetal Bmp2$^{-/-}$; Bmp6$^{-/-}$ mice exhibit persistence of a massive cartilagenous callus in fracture healing. In these models, BMP-6 was found to be exclusively expressed in hypertrophic chondrocytes, and BMP-2 was strongly expressed in hypertrophic chondrocytes and marginally in the osteoblasts, probably because of reduced bone formation and/or stimulation of bone resorption. It may be due to the fact that osteoblasts require endogenous BMPs to assume full function in vivo, yet their origin is unclear. Proposed theories may be that they arise from hypertrophic chondrocytes, that they come from bone and bone marrow, or from other cell sources at a distance. It is apparent that together, BMP-2 and BMP-6 are vital for in-vivo bone formation in both physiological and pathological conditions (Kugimiya et al., 2005).

During distraction osteogenesis, these processes are upregulated, and temporal patterns are different due to the involvement of mechanical stimuli during bone repair. This highlights an additional role for BMPs, which may trigger pathways linking mechanical forces to cellular responses (Rauch et al., 2000).

BMPs also stimulate the expression of vascular endothelial growth factors (VEGFs) and their receptors indicating a linked functional relationship between these two families of growth factors. Angiogenesis plays a critical role during bone growth and repair (Kanczler and Oreffo, 2008). The importance of vascularization is an understatement especially in challenging situations such as critical sized bone defects following tumor resection or trauma.

TGF-β also plays an important role during endochondral bone formation, and its increased concentration appears to be correlated with angiogenesis and calcification of cartilage (Ripamonti and Reddi, 1992). TGF-β2, and TGF-β3 expressions peak on day 7 of fracture healing in the mouse along with high levels of collagen II expression indicating cartilage formation, while TGF-β1 expression remains constant throughout the healing period (Al-Aql et al., 2008).

Angiogenic growth factors guide the formation of new blood vessels during fracture healing, which is mandatory for the transition from the cartilage stage to replacement by bone (Lin et al., 2008; Al-Aql et al., 2008). VEGF-A has been found to mediate vascularization in addition to ensuring normal differentiation of progenitors into hypertrophic chondrocytes, osteoblasts, endothelial cells, and osteoclasts, thereby contributing to endochondral bone formation (Dai and Rabie, 2007).

Angiopoietins are also associated with the formation of blood vessels, and angiopoietin-1 expression is apparent during the initial stages of repair process. On the other hand, platelet-derived growth factor (PDGF) stimulates osteoblast migration and proliferation, and it is also secreted from osteoclasts; therefore, they have a role in bone remodeling as well. In fact, platelet derived growth factors (rhPDGF-BB) have been found to improve alveolar ridge augmentation, in both preclinical and case report studies (Cardaropoli, 2009; Nevins et al., 2009). Understanding the coupling of angiogenesis and osteogenesis in bone regeneration, can aid in designing novel strategies for treatment. Molecular interactions that take place within the cross-talk between endothelial populations and bone populations may indicate the need to consider combinational delivery vehicles capable of providing both angiogenic and osteogenic growth factors to the site of injury (Al-Aql et al., 2008; Kanczler and Oreffo, 2008).

In extraction socket healing, endochondral ossification has not been demonstrated, although chondrocyte related gene expression profiles have not been studied in extraction sites. However, it is assumed that similar growth factors may be present as those found during fracture healing (Kanyama et al., 2003). Although the alveolar bone regenerated using rh-BMP-2 in induced periodontal defects in rats appears to develop intramembranously, yet, endochondral ossification on the original cortical bone adjacent to the defects has been observed (Nakajima et al., 2008). There are obvious spatial and temporal patterns that probably contribute to these events, in addition to tension forces during the healing process, yet, the exact mechanisms governing the pathways of ossification remain enigmatic (Nakajima et al., 2008; King et al., 2007).

Critical cellular events taking place during the healing process of the tooth extraction socket are carefully controlled by signaling molecules such as TGF-β1, BMP-2 and VEGF, which are secreted by the cells recruited to the extraction site to initiate the angiogenic and osteogenic cascades. TGF-β1 expression is evident in the rat extraction socket treated with simvastatin after 5 days. After one week,

BMP-2 and VEGF can be detected, and the expression of all 3 growth factors peaks at two weeks. TGF-β_1 recruits osteoprogenitors and enhances their proliferation as well as enhancing VEGF expression, which triggers angiogenesis and is known to regulate osteoblastic activity in addition to recruitment of mesenchymal progenitors and stimulating their differentiation. BMP-2 induces differentiation of mesenchymal cells to osteoprogenitor cells, osteoblasts and synthesis of collagen by fibroblasts as well as production of proteins of the bone matrix by osteoblasts (Mendes et al., 2008; Liu et al., 2009). Osteopontin (OPN) has also been identified in extraction socket models as it elicits bone matrix mineralization. Maximum expression of OPN was witnessed at days 3 and 7 of healing (Mendes et al., 2008).

Other signaling factors for angiogenesis have also been displayed in healing of extraction sockets such as connective tissue growth factor (CTGF), which is a signaling factor associated with angiogenesis. In fact, CTGF positive endothelial cells have been found in the early healing stage of the extraction socket in rats. In addition, osteoblast-like cells expressing CTGF have also been found (CTGF positive endothelial cells have been found to be expressed by osteoblast-like cells in later stages of healing (Kanyama et al., 2003).

Localization of growth factors in the healing extraction socket can thus provide valuable knowledge into the mechanisms governing these events. BMP-7 positive cells have been found to increase in density from 2-4 weeks to 6-8 weeks and tend to decrease between 12-24 weeks after extraction. The increase in BMP-7 between early and intermediate healing phases explains an increased modeling/remodeling phase, from organization of the provisional matrix to formation of woven bone. This is consistent with studies in fracture healing models which show high expression of BMP-7 during the early stages of fracture bone repair (Trombelli et al., 2008).

It is likely that alveolar bone regeneration necessitates the interplay of several bone morphogenetic proteins residing in the natural extracellular matrix of the periodontal unit. Naturally derived bovine bone morphogenetic proteins such as hTGF-β3 and hOP-1 have been shown to induce alveolar bone and cementum in furcation defects in the Papio ursinus non-human primate. The hTGF-β3 is an important regulator in angiogenesis and osteogenesis. When combined with hOP-1, an enhancement of the endochondral pathway becomes apparent (Ripamonti et al., 2008, 2009).

A wide panel of growth factors has been explored for periodontal tissue engineering and alveolar bone augmentation. These include but are not limited to PDGF, IGFs, FGF, TGF-β, and BMPs (Kitamura et al., 2008; Schwarz et al., 2008; Kinoshita et al., 2008; Scheller et al., 2009) and those that regulate the epithelial-mesenchymal interactions involved in initial tooth formation (e.g., Emdogain™) (Del Fabbro et al., 2009). This is in addition to platelet rich plasma concentrations which have recently gained popular use in clinical applications.

Platelet rich plasma (PRP) represent a pool of growth factors capable of facilitating soft and osseous tissues healing (Anitua et al., 2001; Del Fabbro et al., 2009), although controversy still exists regarding the effectiveness of various preparations. One of the most common reservoirs of growth factors, which has been used experimentally as well as clinically for years, is platelet rich plasma and its derivatives (Intini, 2009). Platelets have granules that when activated release a cocktail of growth factors such as epidermal growth factor, PDGF-A and B, TGF-β1, IGF-1 and 2, vascular endothelial growth factors, and fibroblast growth factors as well as others. These factors interact with cells and promote wound healing of soft tissues and bone (Zaky and Cancedda, 2009; Simon et al., 2009).

Traditionally, platelet rich plasma preparations (PRP) have utilized thrombin for activation, resulting in an immediate release of contained growth factors, thus probably contributing only to the immediate stages of wound healing. Conversely, a platelet rich fibrin material has been developed which does not require a thrombin activator. Platelet-rich fibrin matrix (PRFM), can release up to 6 growth factors sustainably over 7 days. Using PRFM in an extraction socket model in the dog revealed complete osseous filling after 3 weeks while sockets treated with Demineralized Freeze-Dried Bone Allograft (DFDBA) did not demonstrate complete filling even after 12 weeks (Simon et al., 2009).

2.6.2.2 BMPs: From Filling Bone Defects to Enhancing Osseointegration

To date, BMP-2 and BMP-7 (OP-1) are the two members of the BMP family that have received FDA approval for clinical use especially, to supplement fracture repair and spinal fusion (Axelrad and Einhorn, 2009). In human alveolar bone defects, applying rhBMP-2 lyophilized to xenogenic bone substitute Bio-Oss demonstrated a statistically significant enhancement of vertical defect reduction when compared with Bio-Oss alone (Scheller et al., 2009).

Critical-size supraalveolar peri-implant defects treated with rhBMP-2 showed considerable alveolar bone augmentation and enhanced osseointegration within 8 weeks when used in conjunction with an absorbable collagen sponge. Furthermore, when this model was supported by placement of a space-providing macro-porous (ePTFE) GBR device to preclude compression of the construct, vascularity was allowed along with effective bone formation (Wikesjö et al., 2009). Cells transfected with BMP-2 cDNA; therefore, allowing its expression and release, can also significantly improve peri-implant bone regeneration (Lutz et al., 2008). In general, studies utilizing BMP-2 appear to have the most consistent results, as pre-clinical and clinical reports corroborate its ability to enhance bone formation in local augmentation sites (Jung et al., 2008).

An exciting arena for the use of growth factors for alveolar bone engineering is the enhancement of osseointegration of dental implants (Zhang et al., 2007). Alveolar ridge augmentation procedures have long been performed to improve the quality and quantity of bone prior to implant placement. Guided bone regeneration techniques have been clinically implemented for years, Moreover, they have been augmented with BMP-2 delivery in patients (Jung et al., 2009).

The dental implant situation can be further complexed by the growing demand to immediate implants insertion, following extraction in order to prevent alveolar bone resorption and improve patient satisfaction. In this case, there are usually inconsistencies between the size and shape of the implant and the extraction site thereby, warranting the need to assure better osseointegration to guarantee long-term implant survival (Lutz et al., 2008).

Collectively, the previous observations have lead to the conceptualization of the hypothesis to use osteoinductive implants by coating them with bone inductive factors such as rhBMP-2 to stimulate local bone formation and improve osseointegration. Such implants when placed ectopically in rodents or orthotopically in dogs displayed remarkable ability to induce local bone formation, at some instances exceeding the implant platform and sometimes compromising the stability of the implant. Bone-to-implant contact was also improved. Effectively, porous titanium oxide implant surfaces can serve as successful carriers of rhBMP-2 and hOP-1 (BMP-7) to enhance immediate implant osseointegration (Leknes et al., 2008a; Wikesjö et al., 2008; Leknes et al., 2008b).

2.6.2.3 Delivering Growth Factors for Alveolar Bone Engineering

The delivery of growth factors to the target site is one of the obstacles to overcome if growth-factor based strategies are to succeed. It is of optimum importance if the tissues are to retain a substantially high level of the factors to elicit a response in the surrounding microenvironment. BMPs delivered in formulation buffer accelerate bone healing in small animal models, yet, these effects are not so clear in larger animals. Utilizing injectable or implantable carriers to deliver BMPs may allow sufficient volume retention for a long enough time in the treatment site to allow cell recruitment and infiltration, thereby facilitating the healing process (Seeherman and Wozney, 2005).

Modes of delivery include those using carriers that provide sustained release of the growth factor(s), using plasmids (DNA)/vector that include the gene encoding the desired growth factor, transfecting cells with this vector (Chen et al., 2008), or by applying gene therapy principles utilizing non-viral vectors (Sawada et al., 2009).

Carrier requirements range from having to be biocompatible and bioresorbable to displaying a range of porosity that can encompass both the released growth factors and allow space for the cells to infiltrate. This is in addition to maintaining adequate release profiles at the site of delivery. The four major categories of BMP carrier materials include natural polymers such as collagen, hyaluronans, fibrin, chitosan, silk, alginate, and agarose; inorganic materials such as low and high temperature calcium orthophosphates (calcium phosphate cements and sintered ceramics) and calcium sulfate cements, poly(a hydroxy acid) synthetic polymers such as polylactide (PLA), polyglycolide (PLG), and their copolymers (poly(D,L-lactide-co-glycolide) (PLGA). In addition, combinations of natural polymers, calcium orthophosphates, and synthetic polymers have also been evaluated. Allograft and autograft materials have also been used to deliver BMPs (Seeherman and Wozney, 2005). Carriers such as synthetic calcium phosphate cement to deliver rhBMP-2 have been shown to be capable of relevant bone augmentation enhancement (Wikesjö et al., 2002).

Microspheres of gelatin and dextran have been developed as growth factor delivery vehicles for treatment of rabbit alveolar cleft defects, which resulted in earlier bone regeneration and remodeling leading to defect closure (Sawada et al., 2009). Polyurethane scaffolds encoding BMP-4 have also been shown to enhance remodeling of the alveolar bone in induced periodontal defects in an osteoporotic rabbit model (Marei et al., unpublished data).

Other attempts aiming to optimize BMP-2 delivery have examined the osteogenic effects applying BMP-2 cDNA to local cells, which would then begin to produce and secrete BMP-2 protein. A vector consisting of BMP-2 cDNA plasmids and cationic liposomes was combined with collagen or autologous bone as a carrier and delivered to peri-implant bone defects in pig calvaria. Results showed that osseointegration was accelerated with increased bone to implant contact when either of the two carriers was used as compared to control groups that did not receive BMP-2. Bone regeneration in the peri-implant defect was significantly improved although after 28 days defects had not yet totally regenerated. This study sheds light on the prospects of using gene delivery to harness cellular responses, yet, optimal delivery concentrations and release profiles are still to be determined (Lutz et al., 2008).

2.6.3   THE SCAFFOLD: THE ENGINEERED GUIDED TISSUE MIMETIC

In the body, nearly all tissue cells reside in an extracellular matrix (ECM) consisting of a complex 3/D fibrous meshwork with a wide distribution of fibers and gaps that provide complex biochemical and physical signals that influence cell proliferation and differentiation or morphogenesis, which contributes to the resultant tissue regeneration and organogenesis (Lee et al., 2008; Nicodemus and Bryant, 2008; Yang and El Haj, 2006).

ECM is assembled from components synthesized and deposited outside the cell surface that provide structural and functional integrity to connective tissues and organs. The synthesis and deposition of ECM largely occur in response to growth factors, cytokines, and mechanical signals mediated via cell surface receptors (Lodish et al., 2002). To engineer functional bone tissues, cells must be provided with appropriate spatial and temporal cues to enable growth, differentiation and synthesis of an ECM of sufficient volume and functional integrity (Scheller et al., 2009). Bone tissue has a hierarchical organization over length scales that span several orders of magnitude from the macro-(centimeter) scale to the nanostructured components. Bone ECM comprises both a nonmineralized organic component (predominantly type-1 collagen) and a mineralized inorganic component (composed of 4 nm thick plate-like carbonated apatite mineralites) (Stevens, 2008).

Potential materials with suitable characteristics include natural polymers, synthetic polymers, ceramics, metals, and combinations of these materials. (Table 2.3 summarizes the groups of biomaterials used in alveolar bone regeneration). Varying parameters of the biomaterial, such as composition, topology and crystallinity can lead to a significant variation in cell attachment and proliferation, protein synthesis, and RNA transcription in vitro. The parameters of the scaffold can also significantly affect progenitor cell differentiation, amount and rate of tissue formation, and intensity or duration of any transient or sustained inflammatory response in vivo. It was shown that physical nature and chemical composition of the surface either at the atomic, molecular, or higher level relative to the dimensions of the biological units may cause different contact areas with biomolecules and cells, which may influence conformation and function (Scheller et al., 2009; Chou et al., 2005; Chen and Jin, 2010).

The ideal scaffold for periodontal as well as for bone tissue engineering should meet the following requirements: biocompatibility, osteoconductivity, osteoinductivity, bioactivity, high porosity, and an adequate multi-scale pore size is necessary to facilitate cell seeding and diffusion throughout the whole structure of both cells and nutrients. Biodegradability is often an essential factor since scaffolds should preferably be absorbed by the surrounding tissues without the necessity of a surgical removal. The rate at which degradation occurs has to coincide as much as possible with the rate of tissue formation. The scaffold should be able to provide structural and mechanical integrity within the body. Injectability is also important for easy clinical uses (Lee et al., 2008; Scheller et al., 2009; Chen et al., 2007a).

Ceramic scaffolds have been used specifically for alveolar bone repair because they have outstanding properties which include similarity in composition to bone mineral, bioactivity (ability to form bone apatite-like material or carbonate hydroxyapatite on their surfaces), ability to promote cellular function, and expression leading to the formation of a uniquely strong bone-biomaterial interface. It was demonstrated that with appropriate 3D geometry, ceramic materials may bind and concentrate endogenous circulating bone morphogenetic proteins to function as osteoinductive materials. Results of these studies showed that discs of HA with BMPs are osteoinductive, whereas granules are consistently feeble in bone induction, even though their chemical composition and pore size are identical the cellular and molecular mechanisms underlying the role of the geometry of the scaffold with identical chemistry is unclear (Nakashima and Reddi, 2003; Ripamonti et al., 1992; Lu et al., 2001).

Santos et al., 2010 evaluated tissue reaction to two different HAs (synthetic and natural) and bioglass (BG) when implanted into fresh extraction sockets in dogs. The qualitative analysis showed that bovine bone mineral had the highest number of particles involved by bone tissue. Synthetic HA had similar results of bovine bone mineral (natural HA) and showed particles involved with newly formed bone, while other particles were involved by fibrous connective tissue. BG had granules involved by a thin calcified tissue. In some specimens, particles were not observed. None of the tested biomaterial was completely resorbed but showed different rates of resorption. While the materials analyzed showed similar histological characteristics, the BG had the highest reabsorption rate. It was reported that all biomaterials retarded the extraction socket healing (Santos et al., 2010).

De Coster et al., 2009 reported that human socket supplemented with composite of hydroxyapatite (HA) and 40% β-tricalcium phosphate (β–TCP), which sintered at temperatures of 1,100 to 1,500°C with 90% porosity, yields excellent results with respect to volume preservation of the alveolar crest but influences bone healing negatively when compared with naturally healed sockets after equal healing periods (De Coster et al., 2009).

Many synthetic resorbable polymers (e.g., polyglycolide, polylactide, polylactide coglycolide) have been developed to overcome the problems associated with natural polymers. To our knowledge, there are very few number of examples whereby *in vivo* alveolar regeneration has been achieved in fresh extracted socket using scaffolds in combination with cells. Marei et al., 2005 applied the principles of tissue engineering to regenerate alveolar bone in defects created by tooth extraction in rabbits in order to utilize the repaired areas for future reconstruction. A three-dimensional (3-D) hollow root-form porous scaffolds constructed from biodegradable PLA:PGA (50:50) with optimum mechanical characteristics cultured with undifferentiated mesenchymal stem cells. The scaffolds were implanted immediately in defects created by tooth extraction and evaluated radiographically and histologically. At 4 weeks, a dense calcified bone matrix was shown growing with a well organized, architectural pattern of trabecular bone. The alveolar ridge form was preserved without any demarcation between the tissue-engineered implant and new bone formed. The radiographic examination demonstrated that the extraction sockets that received the PLG/cells for 4 weeks were protected against crestal bone reduction, and the bone density measurement of the coronal area of the grafted sockets was almost comparable to the neighbouring natural tooth (Marei et al., 2005).

Naturally, cells, ECM, and growth factors interact differently in bones that have different embryological origins. Therefore, different bones have different trabecular organization, which leads to different functional adaptation. To design a scaffold that is able to functionally mimic the alveolar bone, it is important to understand the mechano-biology of the alveolar bone; the peculiarity of the alveolar bone is due to being under continuous periodontal pressure and tension. The new biomimetic paradigm of tissue engineering requires scaffolds that balance temporary mechanical function with mass transport to aid biological delivery and tissue regeneration. Little is known quantitatively about this balance as early scaffolds were not fabricated with precise porous architecture. Recent advances in both computational topology design (CTD) and solid free-form fabrication (SFF) have made it possible to create biomimetic scaffolds with controlled architecture in terms of chemical composition and physical structure as well as predictable degradation behavior (Chen et al., 2007a; Lutolf and Hubbell, 2005; Hollister, 2005).

Biomimetic scaffolds should be designed such that they elicit specified cellular responses mediated by interactions with scaffold-tethered peptides from extracellular matrix (ECM) proteins, essentially, the incorporation of cell-binding peptides into biomaterials via chemical or physical modification. Such peptides include both native long chains of ECM proteins as well as short peptide sequences derived from intact ECM proteins (Shin et al., 2003). The modification of scaffolds with these peptide sequences can facilitate cellular functions such as adhesion, proliferation, and migration. However, immobilization of growth factors into biomaterials presents one of the easiest methods to achieve such a purpose and has already provided a tangible effect in most experimental and preclinical research (Chen et al., 2007a). Many attempts are done to develop biomaterials for both drug delivery and tissue engineering applications. Growth factor delivery provides a more general and versatile basis for advanced combination device strategies (Hollister, 2005).

Hydrogels present the most desirable candidates for achieving both advantages of intelligently controlled therapeutic proteins delivery and mimicking the structure and biological function of native extracellular matrix (ECM) for cell proliferation and differentiation as much as possible, both in terms of chemical composition and physical structure. Maire et al. recently reported the fabrication of a kind of functionalized dextran-derived hydrogels loaded with bone morphogenetic protein (BMP) for bone healing enhancement, but, the in vitro release kinetics indicated that hydrogels with a very rapid burst release manner could only retain BMP for a few days, which is, obviously, far from our expectation (Maire et al., 2005). The pharmacokinetics of the BMPs, in conjunction with biomaterials, may be different from kintics of the alveolar bone because retention of BMPs at the site of implantation is dependent on the charge characteristics and isoelectric point of the morphogenes (Nakashima and Reddi, 2003).

Chen et al. developed novel hybrid biodegradable macroporous hydrogel scaffolds that synthesized from glycidyl methacrylated dextran (Dex-GMA)/gelatin scaffolds containing microspheres loaded with bone morphogenetic proteins. Such hydrogel-microsphere composition scaffolds, with a macroporous and interconnected pore structure, were confirmed to have the potential to enhance the osteogenic differentiation of periodontal ligament cells in vitro and accelerate periodontal tissue regeneration in dogs (Chen et al., 2007a).

Hyaluronan (HY) based gels have been used in bone defects, alone or associated with bone morphogenetic protein (BMP-2), demineralized bone matrix, hydroxyapatite, or bone graft. Mendes et al. demonstrated that HY accelerated the healing process of rat sockets after extraction of upper first molars. HY treated sockets showed a more pronounced bone deposition and vessels formation and a less amount of cells. The higher amount of blood vessels could be due to the ability of the HY to stimulate the migration and proliferation of endothelial cells, accelerating the neoformation of blood vessels and, consequently, the bone deposition. During the healing process, there is a progressive replacement of blood vessels and inflammatory cells by bone by stimulating the expression of osteogenic proteins. It has been suggested that the therapeutic effects of the HY depend on its molecular weight and small chains of HY are therapeutically more efficient than chains with high molecular weight (Mendes et al., 2008).

The fragile nature of proteins has motivated the design of scaffolds to release naked plasmid DNA containing genes that encode growth factors (Storrie and Mooney, 2006). Zhang et al. investigated the influence of a chitosan/coral composite scaffold combined with human PDL cells and fabricated to release the gene for platelet-derived growth factor (PDGF) and implanted subcutaneously into athymic mice. Results indicated that PDL cells showed enhanced proliferation properties on the gene-activated scaffolds as compared to pure coral scaffolds, and the expression of PDGF and type-I collagen was up-regulated in the gene-activated scaffold. Following *in vivo* implantation, PDL cells not only proliferate but also increase their expression of PDGF. In another study from the same group, porous chitosan/collagen scaffolds were prepared through a freeze-drying process and loaded with an adenoviral vector encoding BMP-7. Results indicated that the scaffold containing Ad-BMP-7 exhibited higher alkaline phosphatase activity and that the expression of osteopontin and bone sialoprotein were up-regulated. After implantation in defects around implants, bone formation in Ad-BMP-7 scaffolds was greater than that in other scaffolds at 4 or 8 weeks, demonstrating the potential of chitosan/collagen scaffolds combined with Ad-BMP-7 in bone tissue engineering (Zhang et al., 2007).

Based on current evidence from protein-based therapies, protein/gene delivery scaffolds suitable for implantation at load-bearing sites offer a greater degree of versatility for clinical applications (Zhang et al., 2003). The control of gene expression by cells within a scaffold can be regulated *via* interactions with the adhesive surface, with other cells in the vicinity, or with growth and differentiation factors incorporated into the scaffold. Accordingly, cell-seeding scaffolds must provide the correct combination of these factors, depending upon the nature of the tissues to be regenerated (Ramseier et al., 2006). To date, little work has been carried out in this complex area, although early studies have begun to utilize specific cell-attachment peptide sequences, pore sizes, and surface textures in an attempt to improve tissue integration and regeneration. Controlling the diffusion rates of genes and proteins from scaffolds so that they are within the physiological range is the next challenge. New bioactive materials, such as those that covalently incorporate growth factors and other molecules that regulate cell behavior, offer alternatives for enhancing scaffold performance (Chen and Jin, 2010; Chen et al., 2009).

2.7   CURRENT PROMISING APPLICATION

Current scientific effort seeks to solve the still problematic treatment of large bony defects in oral and craniomaxillofacial surgery prior to oral rehabilitation. These defects remain serious problem as the associated loss of function and esthetic considerably impairs the quality of life of the affected patient. The existed techniques do in general result in a favorable clinical outcome, but it is associated with undeniable drawbacks such as a considerable increase in surgical procedures, as well as elevate postoperative morbidity to the patient and connectedly high costs to the socio-economic system (Schieker and Mulschler, 2006; Drosse et al., 2008).

Since the introduction of tissue engineering into the scientific community 20 years ago, several efforts have been undertaken to transfer the technology from research to beside. Alveolar bone defects resulting from tooth loss, trauma, congenital abnormalities, progressive deforming diseases or oncological resection, present a formidable challenge and restoration of these is a subject of clinical, basic science, and engineering concern (Miiller et al., 2007; Lopez-Roldan et al., 2009). There are several locations in the oral/craniofacial region where repair, regeneration of alveolar bone is vital, e.g., the alveolar defect results from extraction of the tooth. In this condition, regeneration of this defect usually results in preservation of the remaining buccal and lingual walls of the socket. This wound, should regenerate alveolar bone till the crestal level of the remaining socket walls, so the end result is alveolar ridge that can receive dental implant and preserve the distance between the two neighboring teeth to the extracted socket and prevent these neighboring teeth from tilting and reduce the existed space for rehabilitation.

Within this context, there are numbers of multi-component approaches that are being developed in tissue engineering alveolar bone in the socket area.

The earlier work of "Lekovic et al., 1998" utilizing the principle of guided bone regeneration via bioabsorbable membrane made of glycolide and lactide polymers with or without bone grafting that demonstrated alveolar ridge defects prevention in clinical trials have opened a vision and strategy for application of tissue engineering in this area (Lekovic et al., 1998). Preserving alveolar ridge was shown to be achieved by providing cell line and metabolically active cells that are able to repair

Table 2.3: Candidate scaffolding materials for alveolar bone regeneration. (*Continues*).

Authors	Scaffolding Materials	Study Model	Results
Hoffmann et al 2008	high-density polytetrafluoroethylene (dPTFE) membranes without the use of a graft material	Human extraction sockets	The use of dPTFE membranes predictably led to the preservation of soft and hard tissue in extraction sites.
Dekok et al., 2005	hydroxyapatite/ tricalcium phosphate (HA/TCP) cylinders	Extraction sockets in beagle dogs	Local bone repair occurred in the absence of nonspecific differentiation or migration with distant osteogenesis.
Wikesjö et al., 2008	Titanium implants coated with rhBMP-2	Alveolar ridge defects in dogs	rhBMP-2 coated onto titanium porous oxide implant surfaces induced clinically relevant local bone formation including vertical augmentation of the alveolar ridge and osseointegration. Higher concentrations/doses were associated with untoward effects.
Araújo et al., 2009	Bio-Oss Collagen	Extraction sockets in dogs	The placement of Bio-Osss Collagen in the fresh extraction wound obviously delayed socket healing. Thus, after 2 weeks of tissue repair, only minute amounts of newly formed bone occurred in the apical and lateral borders of the grafted sockets, while large amounts of woven bone had formed in most parts of the nongrafted sites.

Table 2.3: Candidate scaffolding materials for alveolar bone regeneration. (*Continued*).

Authors	Scaffolding materials	Study Model	Results
Mendes et al., 2008	Sodium hyaluronate (HY) gel	Tooth sockets in rats	Histological analysis showed that HY treatment induced earlier trabecular bone deposition resulting in a bone matrix more organized at 7 and 21 days after tooth extraction. Also, expression of BMP-2 and OPN was enhanced in HY-treated sockets compared with control sockets.
Serino et al., 2008	Bioabsorbable polylactide-polyglycolide acid sponge (Fisiografts)	Human extraction sockets	The bone formed 3 months after the extraction was rich in osteoblasts and newly formed blood vessels. The biocompatibility, safety, and characteristics of Fisiografts suggest that the material is suitable for filling alveolar sockets following extractions, to prevent volume reduction and collapse of the overlying soft tissue flaps.
Nakajima et al. 2008	Graft composed of fibronectin (FN) matrix-based multilayered cell sheets of human gingival fibroblasts modified to express alkaline phosphatase (ALP) (FN-ALP)	Alveolar defects in rats	FN-ALP transplants healed alveolar bone defects by intramembranous ossification, with formation of cementum and periodontal ligament. Moreover, FNALP transplants increased new bone formation, by endochondral ossification, on the mandibular cortex adjacent to the defect.

a defect site through their continuous matrix synthesis. Pre-clinical in-vivo studies have proven generally positive effect of live cells on bone regeneration, particularly, the osteogenic potential of mesenchymal stem cells. Hydroxyapatite/tricalcium phosphate shaped in a cylinder form scaffold and seeded with bone-marrow derived mesenchymal stem cells induced alveolar bone regeneration in 21 days (Dekok et al., 2005; Yang et al., 2000).

The diversity of required characteristics of the scaffold was shown to couple with need for the tissue engineered alveolar bone to withstand forces of mastication. It become clear during the last two decades that biomechanical consideration is fundamental to the success of engineering living functional tissues with high strength and endurance (Engler et al., 2006; Mauck et al., 2007; Butler et al., 2009).

The scaffold prepared by this technique provided support to the crestal boney wall of the socket while wound healing proceeded (Fig. 2.5 a-c). In addition to the viscoelastic characteristics of these PLG/cell scaffolds under masticatory load, they played a vital role in cell attachment, viability and functionality and was shown to recently synergistically enhance osteoblast differentiation and mineralization within the 3/D extracellular matrix (Datta et al., 2006; Engler et al., 2007).

These experiments proven that the socket height, width, and form were maintained; bone density for the healed socket was enhanced in cell/PLG construct graft versus the PLG graft alone. The most important result was that the neighboring teeth maintained their position, keeping the distance for future rehabilitation in human clinical trials (Faramawy, 2006; Marei et al., 2003a) (Fig. 2.5 d,e).

To date, direct growth factor delivery and blank or growth factor containing scaffolds (conductive and inductive materials) are strategies being discussed for alveolar bone regeneration. Potential healing and regeneration was shown in the extraction sockets treated with platelet-rich fibrin matrix at 12 weeks in dog models; the benefits of this technique is to perform ridge preservation without need to use a membrane (Simon et al., 2009; Sammartino et al., 2005).

Considerable scientific effort has been directed towards using natural or synthetic components of extra cellular matrix; these components were shown to stimulate migration, adhesion, proliferation, cell differentiation expression of BMP and osteogenic factors, and vascular endothelial growth factors that resulted in bone formation in the extracted sockets (Mendes et al., 2008; Liu et al., 2009; Lalani et al., 2005; Wu et al., 2008).

Alveolar bone regeneration as a part of periodontal structure therapy was shown to develop when using bone marrow mesenchynal stem cells seeded on PLG or calcium alginate scaffold in extracted sockets. These scaffolds prepared to degrade faster, so they did not maintain long enough to fill the whole area with bone, instead they showed periodontal structure where alveolar bone healed along the internal socket walls while the remaining regenerated tissues formed cementum on implant surface and periodontal fibers in between (Fig. 2.5 f, g) (Marei et al., 2009; Lalani et al., 2003; Weng et al., 2006).

In the past 20 years, the number of dental implant procedures have increased steadily, reaching about one million dental implants per year (Chiapasco et al., 2006; Marco et al., 2005). Immediate

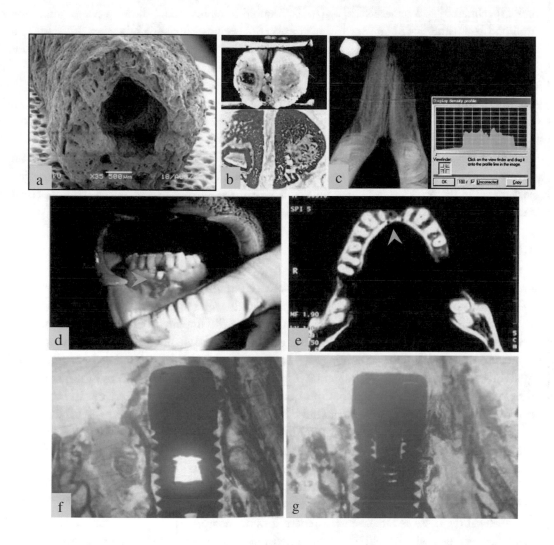

Figure 2.5: *Caption on next page.*

Figure 2.5: *Caption for figure on previous page.* Alveolar bone regeneration via Tissue Engineering (Plate A): (a) PLG hollow root form scaffold seeded with bone marrow derived mesenchymal stem cells prior to implantation in extracted socket. (b) 4 weeks post implantation, macroscopic, and microscopic view for bone specimen of sockets received PLG/cells. Scaffold in rabbit model showing preservation of socket wall and regeneration of bone formation from periphery toward center of socket, X40. (c) A digital radiograph and bone mineral density for lower incisors teeth in rabbit model 4 weeks post implantation demonstrating the preservation of the socket wall and the distance used to be occupied by the extracted tooth (Courtesy of Marei et al. with permission). (d,e) Clinical human trials: In (d) represents clinical photo for immediate implantation of hollow root form PLG scaffold after tooth removal; clinical Trials 1. In (e) CT image for the hollow root form scaffold in place, demonstrating the preservation of the socket wall; clinical trials (Marei et al. patent #23731/2003, Egypt). (f) Implant fixture inserted immediately after tooth extraction and surrounded by BMSCs/PLG scaffold at the same time; goat model (g) Implant fixture inserted with PLG scaffold around it with no cells. Notice alveolar bone regeneration along the socket wall in the first case and along the whole side of the implant fixture, which did not occur in the second case (Courtesy of protectMarei et al., 2009).

implantation after tooth extraction is an attractive alternative that presents several advantages, such as reduction of post extraction resorption, presentation of socket wall height and alveolar ridge form and width, in addition to optimal positioning of the implant. While sufficient bone volume is an implant prerequisite for dental implant placement and osseointegration, preserving the shape, form, and dimension of the remaining alveolar ridge to obtain the expected esthetic and functional goals are all vital for the success of dental implant therapy. Definitive patterns of mineralized tissue development during osseointegration and bone remodeling are considered the indicators for the survival rate of the state-of-the-art dental implantology (Lin et al., 2010; Jung et al., 2008).

Calcium phosphate materials were used to induce bone tissue engineering that was later transplanted to support dental implant placed in mandibular defect (Yao et al., 2009, (22)). Bone morphogenetic proteins BMP-2 with demineralized bone powder or calcium phosphate carriers was also shown to induce significant alveolar ridge augmentation in dogs and revealed excellent clinical and radiological outcomes after 3 and 5 years in patients (Jung et al., 2009).

Recently, there is increased awareness toward incorporation of remodeling laws to various implants settings, which is believed to contain significant potential impact on futuristic dental implants, not only to preserve extraction socket walls, but also to permit remodeling of the surrounding alveolar bone (Chou et al., 2008; Li et al., 2007) and provide aesthetic dental implant treatment (Yamada et al., 2008).

One of the very promising application of tissue engineering in the oral cavity is the alveolar cleft bone regeneration. It offers several advantages, e.g., improvement of oral hygiene, nasal support, closure of Oro-nasal fistula, bone support for dental implants, and maxillary arch continuity for orthodontic treatment. These therapeutic advantages enhance quality of life for cleft patient (Rawashdeh and Telfah, 2008; Yen et al., 2005) (Fig. 2.6 a-d).

The multi-component strategies of tissue engineering were shown to be promising in this area of bone engineering. One potential approach is to harvest stem cells from small fragments of orbicularis oris muscle that are regularly discarded during the first surgical procedure performed in the rehabilitation of cleft palate patients. These cells are capable of osteogenic differentiation and are shown to induce bone regeneration when implanted with a collagen membrane in critical size cranial defects (Bueno et al., 2009, (31)). Another alternative strategy would be to utilize bone marrow stem cells/BMP-2 supplied with scaffolds such as collagen sponge, platelet-rich plasma, or gelatin hydrogels, which have shown promising results in preclinical and clinical trials (Hibi et al., 2006; Gimbel et al., 2007; Dickson et al., 2008; Sawada et al., 2009).

Enlarged maxillary sinus and reduced residual bone height in the posterior maxilla remain a major challenge in modern implant dentistry. Sinus floor augmentation became a widely accepted and a routine method to improve the amount of bone volume before/during implant placement (Tiwana et al., 2006; Browaeys et al., 2007; Kirmeier et al., 2008).

Presently, autogenous bone grafts are considered the "gold standard" although their disadvantages (low availability of intra-oral bone tissue, high surgery costs, and post operative morbidity) have stimulated the search for alternative sources, e.g., various osteoconductive and materials have been proposed (Browaeys et al., 2007).

In cases of sever atrophy of the maxillary alveolar process, sinus floor elevation and implants insertion are usually performed in two stages. When autogenous bone graft bone graft is used, it takes approximately six months following augmentation for the transplanted bone to be integrated and substituted by osteoconduction (Cordaro, 2003; Wood and Moore, 1988).

Alternatively, autogenous bone transplanted was replaced by many investigations with bone substitutes, e.g., freeze dried bone (FDB) allografts; however, maturation of these materials was reported to take up to eight months when used for sinus floor augmentation (Kassolis et al., 2000). Although, it was reported that FDB was not suitable to replace autogenous graft (Haas et al., 2002), other investigations were able to demonstrate new bone formation when used alone or in combination with other materials in clinical trials (Schwartz et al., 2007). Identification of appropriate scaffolds, cell sources, and spatial and temporal signals are necessary to optimize development of new bone interface and functional requirements. The impact of scaffold characteristics, its strategic design seeded with cells, growth factors, or combination of all is still a controversial issue, in the area of tissue engineering application for sinus augmentation. Platelet-Rich Plasma (PRP) and bone-marrow derived from stromal cells were shown to enhance bone formation and increased bone volume in rabbits at four and eight weeks (Ohya et al., 2005), reduce discomfort with minimal invasive technique, and predict implant success in human trials after six years (Yamada et al., 2008). Other findings indicated that PRP combined with HA/fluoro-hydroxyapatite is considered more satisfactory in maxillary sinus grafting than any of these ingredients alone (Fürst et al., 2003; Pieri et al., 2008).

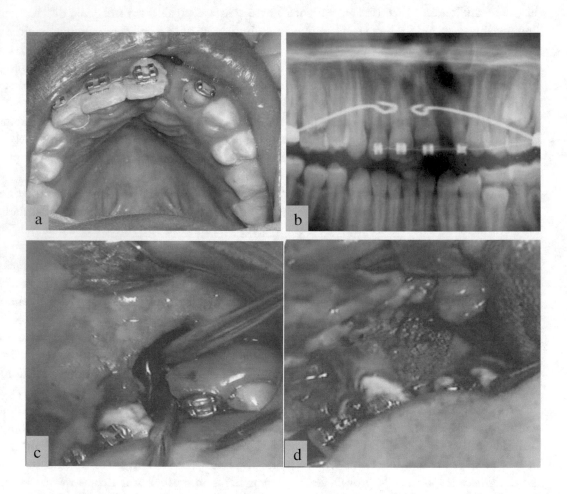

Figure 2.6: *Figure continues on next page.*

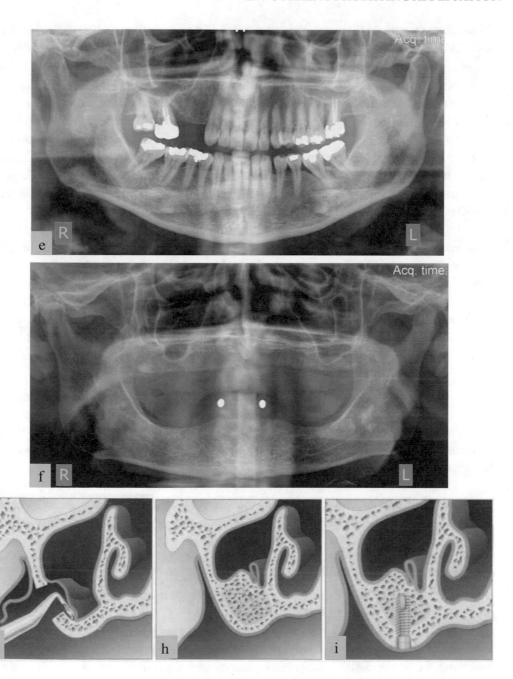

Figure 2.6: *Figure continues on next page.*

Figure 2.6: *Caption for figures on previous pages.* Alveolar bone grafting via Tissue Engineering (Plate B): (a-d) Cleft lip and palate patient (11 years old) has an alveolar bone defect in the left side operated for secondary alveolar bone grafting using autologous bone from the iliac crest (coffin lid technique) after orthodontic maxillary expansion to provide alveolar bone for orthodontic realignment of the left lateral and canine. (a) Macroscopic photograph for the defected area before a secondary alveolar bone grafting. (b) Orthopantomogram x-ray showing the bone defect preventing teeth realignment. (c) The oral cavity clinically showing the extent of the alveolar bone defect on the left side. (d) Filling the alveolar defect with autogenous bone graft operated from the iliac crest. (Courtesy of Dr. Amr M. Osama, Oral and Maxillofacial Department, Faculty of Dentistry, Alexandria University, Egypt.) (e,f) Orthopantomogram Radiography for a patient showing sinus geometry in relation to alveolar ridge in maxilla; (e) in relation to roots of maxillary posterior teeth, (f) in case of missing teeth.

(g-h) Illustrative diagrams showing maxillary sinus augmentation grafting technique in human using a modified window shape and design. (g) Coronal view of sinus membrane elevation (before grafting). (h) Sinus cavity grafted with the amount of material needed for future implant placement. (i) Coronal view of the implant site after implant in placed and surrounded by sufficient bone. (Illustrative drawings are from Garg AK. Augmentation Grafting of the Maxillary Sinus for Placement of Dental Implants. In: Bone Biology, Harvesting, Grafting for Dental Implants: Rational and Clinical Application. 2004 Quintessence Pub. pp. 171–211.) (reprinted with permission from Quintessence Publishing Co Inc.)

So, the use of Platelet-Rich Plasma was a promising option, yet it remains controversial. Platelet-Rich Fibrin (PRF) offers a new generation of platelet concentration that enhances the performance of Freeze Dried Bone Allograft (FDBA) for bone regeneration in sinus floor elevation. This combined effect of PRF and FDBA is believed to reduce the healing time from 8 to 4 months in human clinical trials (Choukroun et al., 2006).

Different cell sources were investigated for cell/scaffold construction in tissue engineering application in maxillary sinus floor augmentation. Although, periosteal cells isolated from lateral cortex of mandibular angle were shown to have an osteogenic behavior upon transplantation. Yet, periosteal cells/polymer fleece did not show enough positive potential for bone engineering in sinus floor augmentation/implant insertion on human clinical treatment (Schmelzeisen et al., 2003; Schimming and Schmelzeisen, 2004). In clinical trials study, some authors felt that autogenous cancellous bone grafts in sinus augmentation was more reliable than osteoblasts/seeded on polyglycolid-polylactic scaffolds (Zizelmann et al., 2007).

While, these previous results were all obtained with polymer scaffold, there were reported data of using BMSCs and hydroxyapatite particles with various rate of success. Several investigations suggested the use of BMSCs/Ceramic/Biphasic HA-TCP/bone mineral scaffolds as a substitute to autogenous bone graft in case of immediate placement of dental implant in maxillary posterior region. These studies showed earlier promotion of bone formation and mineralization, maximally maintain height, volume and, increase compressive strength of augmented maxillary sinus in rabbits up to eight weeks, in dogs up to 20 weeks and in human nine month to seven years (Sun et al., 2008; Wang et al., 2010; Shayesteh et al., 2008; Springer et al., 2006).

In this section of this monograph, we shined light on the alveolar bone as a vital part of oral and maxillofacial structure, and its presence reflecting youth and health, its maintenance as a key to good quality of life. A review of the current concept of tissue engineering to regenerate alveolar bone in various locations, utilizing the multi component approach of the field, was discussed. Tissue engineering is undergoing a major development since it originated 20 years ago with knowledge, effort, and understanding transformation occurred to attempt the *in vitro* processes that mimic original development. What happens with regeneration of any tissue in the body can be applied in the orofacial region, and it can be modified to be suitable for the unique function of this highly demanded structure.

2.8 ACKNOWLEDGMENTS

1. Ministry of Scientific Research; Academy of Scientific Research and Technology - Science and Technology Center - Program of the National Strategy for Biotechnology and Genetic Engineering.

2. US/Egypt science and technology joint fund, Ministry of International Cooperation /with the United States of America for funding the high technology equipment.

3. Alexandria University for supporting infrastructure for material lab., animal surgery, sterilization lab., and goat husbandry.

4. Pharco-pharmaceutical company for establishment the infra structure for the stem cell facility, rabbit husbandry.

5. Faculty of Dentistry – Alexandria University for hosting the tissue engineering laboratories and maintaining them since 1999.

BIBLIOGRAPHY

Akintoye SO, Lam T, Shi S, et al. Skeletal site-specific characterization of orofacial and il-iac crest human bone marrow stromal cells in same individuals. Bone. (2006); 38: 758–768. DOI: 10.1016/j.bone.2005.10.027 40

Al-Aql ZS, Alagl AS, Graves DT, et al. Molecular Mechanisms Controlling Bone Formation during Fracture Healing and Distraction Osteogenesis. J Dent Res. (2008); 87: 107–118. 32, 40, 41, 42, 43

Allori AC, Sailon AM and Warren SM. Biological Basis of Bone Formation, Remodeling, and Repair—Part I: Biochemical Signaling Molecules. Tissue Engineering: Part B. (2008); 14 (3): 259–273. DOI: 10.1089/ten.teb.2008.0082 20

Allori AC, Sailon AM, and Warren SM. Biological Basis of Bone Formation, Remodeling, and Repair—Part II: Extracellular Matrix. Tissue Engineering: Part B. (2008); 14 (3): 275–283. DOI: 10.1089/ten.teb.2008.0083 20

Allori AC, Sailon AM, and Warren SM. Biological Basis of Bone Formation, Remodeling, and Repair—Part III: Biomechanical Forces. Tissue Engineering: Part B. (2008); 14 (3): 285–293. DOI: 10.1089/ten.teb.2008.0084 20

Amler MH. The time sequence of tissue regeneration in human extracted wounds. Oral surgy. (1969); 27: 309–318. DOI: 10.1016/0030-4220(69)90357-0 32, 36

Anitua E, Ardanza B., Paponneau A, et al. (2001) Clots from platelet-rich plasma promote bone regeneration in so doing reducing the need for dental implants and favouring their osteointegration. Blood. 2001; 11: 242a. 44

Anusaksathien O, Jin Q, Ma PX, et al. In: Ma PX and Elisseff J. Editors. Scaffolding in tissue engineering. CRC press, (2006), 437–446.

Araújo M, Linder E, Lindhe J. Effect of a xenograft on early bone formation in extraction sockets: an experimental study in dog. Clin Oral Implants Res. (2009); 20(1): 1–6. 52

Arrington ED, Smith WJ, Chambeuo HG, et al. complications of iliac crest bone graft harvesting. Clin orthop relat res. (1996); 329: 300. 30

Artzi Z, Tal H, and Dayan D. Porous bovine bone mineral in healing of human extraction sockets. 1. Histomorphometric evaluations at 9 months. J. Periodontol. 2000; 71: 1015. DOI: 10.1902/jop.2000.71.6.1015 30

Ashman A. Ridge preservation: important buzzwords in dentistry. Gen.Dent (2000); 38: 304–309. 29

Atwood DA. Postextration changes in the adult mandible as illustrated by microradiographs of mid-sagittal sections and serial cephalometric roentgenograms. J Prothet Dent. (1963); 13: 810–824. 29

Atwood DA. The reduction of residual ridge resorption, a major oral disease entity. J Prosthet Dent (1971); 26: 266–279. DOI: 10.1016/0022-3913(71)90069-2 29

Avery JK and Chiego DJ. Periodontium: Alveolar Process and Cementum. In: Essentials of Oral Histology and Embryology. A Clinical Approach. 3rd edition, Mosby Elsevier Inc., 2006; pp 157–166. 26, 27

Avery JK. Histology of the periodontium: Alveolar bone, cementum, and periodontal Ligament. In: Avery JK. Oral Development and Histology. (2001) 3rd edition. pp 226. 27

Axelrad TW and Einhorn TA. Bone morphogenetic proteins in orthopaedic surgery. Cytokine Growth Factor Rev. (2009); 20(5-6): 481–488. DOI: 10.1016/j.cytogfr.2009.10.003 41, 45

Berkovitz BK, Holland GR, Moxham BJ. Oral anatomy, histology and embryology. Third ed., Mosby Elsevier, (2002); pp 205–219. 27

Bernard GW, Pease DC. An electron microscopic study of initial intramembranous osteogenesis. Am J anat (1969); 125: 271–290. DOI: 10.1002/aja.1001250303 26

Betz RR. Limitations of autograft and allograft: new synthetic solutions. Orthopedics. (2002); 25(5 Suppl): s561–70. 30

Bhaskar SN. Maxilla and Mandible (Alveolar Process). In: Orban's Oral histology and embryology. 11th ed Mosby-Year Book, Inc. 1991. pp 239–259. 23, 24

Bonucci E. Fine structure of early cartilage calcification. J Ultrasruct Res. (1967); 20: 33–50 24

Boyne PJ. Bone grafting: materials. In: Evensen, L., ed. Osseous reconstruction of the maxilla and mandible. Chicago, IL: Quintessence Publishing, (1997) pp. 3–11. 30

Browaeys H, Bouvry P, De Bruyn H. A literature review on biomaterials in sinus augmentation procedures. Clin Implant Dent Relat Res. (2007); 9(3): 166–77. DOI: 10.1111/j.1708-8208.2007.00050.x 57

Brugnami F, Caiazzo A, Leone C. Local intraoral autologous bone harvesting for dental implant treatment: alternative sources and criteria of choice. Keio J Med. (2009); 58: 24–28. 40

Bueno DF, Kerkis I, Costa AM, et al. New source of muscle-derived stem cells with potential for alveolar bone reconstruction in cleft lip and/or palate patients. Tissue Eng Part A. (2009); 15(2): 427–35. 57

Butler DL. Galdstcin SA. Guldberg RE. et al. The impact of biomechanics in tissue engineering and regenerative medicine. Tissue Eng. (2009) part B; 15: 477–484. DOI: 10.1089/ten.TEB.2009.0340 54

Camargo PM, Lekovic V, WeinLaender M., et al. Influence of bioactive glass on changes in alveolar process dimensions after exodontia. Oral Surg. Oral Med. Oral Pathol. Oral Radiol. Endod. (2002); 90: 581. DOI: 10.1067/moe.2000.110035 31

Cardaropoli D. Vertical ridge augmentation with the use of recombinant human platelet-derived growth factor-BB and bovine bone mineral: a case report. Int J Periodontics Restorative Dent. (2009); 29(3): 289–295. 43

Cawood JJ and Harvell RA. A classification of the edentulous jaws. Int J Oral Maxillofac Surg (1988); 7: 232–236. DOI: 10.1016/S0901-5027(88)80047-X 29

Chan CK, Chen CC, Luppen CA, et al. Endochondral ossification is required for haematopoietic stem-cell niche formation. Nature. (2009); 457: 490–494. 40

Charbord P, Garrett RW, Emerson SG, et al. Early ontogeny of the human marrow from long bones: an immunohistochemical study of hematopoiesis and its microenvironment. Blood. (1996); 87: 4109–4119. 39

Chen FM and Jin Y. Periodontal Tissue Engineering and Regeneration: Current Approaches and Expanding Opportunities. Tissue Engineering Part B: Reviews. (2010). In print. DOI: 10.1089/ten.teb.2009.0562 20, 47, 51

Chen FM, Wang Q, et al. Gene Delivery for Periodontal Tissue Engineering: Current Knowledge - Future Possibilities. Curr Gene Ther. (2009); 9(4): 248–66. 51

Chen FM, Zhao YM, Sun HH, et al. Novel glycidyl methacrylated dextran (Dex-GMA)/gelatin hydrogel scaffolds containing microspheres loaded with bone morphogenetic proteins: formulation and characteristics. J Control Release. 2007 12; 118(1): 65–77. DOI: 10.1016/j.jconrel.2006.11.016 48, 49, 50

Chen FM, Zhao YM, Zhang R, et al. Periodontal regeneration using novel glycidyl methacrylated dextran (Dex-GMA)/gelatin scaffolds containing microspheres loaded with bone morphogenetic proteins. J Control Release. 2007; 121(1-2): 81–90. DOI: 10.1016/j.jconrel.2007.05.023

Chen YL, Chen PK, Jeng LB, et al. periodontal regeneration using ex-vivo autologous stem cells engineered to express the BMP-2 gene: an alternative to alveoloplasty. Gene Ther. 2008; 15(22): 1469–1477. 46

Chiapasco M, Abati S, Romeo E, and Vogel G. Clinical outcome of autogenous bone blocks or guided bone regeneration with e-PTFE membranes for the reconstruction of narrow edentulous ridges. Clin. Oral Implants Res. (1999); 10: 278. 30

Chiapasco M, Zaniboni M, Boisco M. Augmentation procedures for the rehabilitation of deficient edentulous ridges with oral implants. Clin Oral Implants Res. (2006) Oct; 17 Suppl 2: 136–59. Review. DOI: 10.1111/j.1600-0501.2006.01357.x 54

Chou HY, Jagodnik JJ, Müftü S. Predictions of bone remodeling around dental implant systems. J Biomech. 2008; 41(6): 1365–73. Epub 2008 Apr 3. DOI: 10.1016/j.jbiomech.2008.01.032 56

Chou YF, Huang W, Dunn JC, Miller TA, Wu BM. The effect of biomimetic apatite structure on osteoblast viability, proliferation, and gene expression. Biomaterials. (2005); 26(3): 285–95. DOI: 10.1016/j.biomaterials.2004.02.030 47

Choukroun J, Diss A, Simonpieri A, Girard MO, Schoeffler C, Dohan SL, Dohan AJ, Mouhyi J, Dohan DM. Platelet-rich fibrin (PRF): a second-generation platelet concentrate. Part V: histologic evaluations of PRF effects on bone allograft maturation in sinus lift. Oral Surg Oral Med Oral Pathol Oral Radiol Endod. 2006; 101(3): 299–303. DOI: 10.1016/j.tripleo.2005.07.012 60

Colnot C. Skeletal cell fate decisions within periosteum and bone marrow during bone regeneration. J Bone Miner Res. (2009); 24: 274–282. 38

Cordaro L. Bilateral simultaneous augmentation of the maxillary sinus floor with particulated mandible. Report of a technique and preliminary results. Clin Oral Implants Res. (2003); 14(2): 201–6. 57

Corelini R, Scarano A, Covani U, et al. Immediate onestage post-extraction implant: A human clinical and histologic case report. Int. J. Oral Maxillofac. Implants (2000); 15: 432. 30

Dai J and Rabie ABM. VEGF: an Essential Mediator of Both Angiogenesis and Endochondral Ossification. J Dent Res. (2007); 86: 937–950. DOI: 10.1177/154405910708601006 43

Dalle Carbonarea L, Valentia MT, Bertoldoa F, et al. Bone microarchitecture evaluated by histomorphometry. Micron (2005); 36: 609–616. DOI: 10.1016/j.micron.2005.07.007 26

Datta N, Pharm QP, Sharm AU, et al.: In-vitro generated extracellular matrix and fluid shear stress synergistically enhance 3/D osteoblast differentiation. Proc. Natl. Acad. Sci. USA. 2006; 103: 2488. DOI: 10.1073/pnas.0505661103 54

De Coster P, Browaeys H, De Bruyn H. Healing of Extraction Sockets Filled with BoneCeramic® Prior to Implant Placement: Preliminary Histological Findings. Clin Implant Dent Relat Res. (2009). [Epub ahead of print.] 48

De Vicente JC, Lopez-Arranz E, and Lopez-Arrans JS. Tissue regeneration in bone defects adjacent to endosseous implants: An experimental pilot study. Int. J. Periodontics Restor. Dent. (2000); 20: 41. 30

Debra L. Wright, Kenneth V. Kardong, David L. Bentley. The Functional Anatomy of the Teeth of the Western Terrestrial Garter Snake, Thamnophis elegans. Herpetologica. (1979); 35 (3): 223–228. 20

Dekok IJ, Drapeau SJ, Young R, et al.:evaluation of mesenchymal stem cells following implantation in Alveolar sockets: a Canine safety study. The Int. Journal of oral and maxillofacial implants (2005); 20: 511–518. 52, 54

Del Fabbro M, Boggian C, Taschieri S. Immediate implant placement into fresh extraction sites with chronic periapical pathologic features combined with plasma rich in growth factors: preliminary results of single-cohort study. J Oral Maxillofac Surg. (2009); 67(11): 2476–84. DOI: 10.1016/j.joms.2009.04.063 44

Deschaseaux F, Pontikoglou C, Sensébé L. Bone regeneration: the stem/progenitor cells point of view. J Cell Mol Med. 2010; 14(1-2): 103-15. DOI: 10.1111/j.1582-4934.2009.00878.x 39

Devlin H and Sloan P. Early bone healing events in the human extraction socket. Int. J. Oral Maxillofac Surg. (2002); 31: 641–645. DOI: 10.1054/ijom.2002.0292 32, 33

Dickinson BP, Ashley RK, Wasson KL, O'Hara C, Gabbay J, Heller JB, Bradley JP. Reduced morbidity and improved healing with bone morphogenic protein-2 in older patients with alveolar cleft defects. Plast Reconstr Surg. 2008; 121(1): 209–17. DOI: 10.1097/01.prs.0000293870.64781.12 57

Dimitriou R, Tsiridis E, Giannoudis PV. Current concepts of molecular aspects of bone healing. Injury. (2005); 36: 1392–1404. DOI: 10.1016/j.injury.2005.07.019 42

Dominici M, Pritchard C, Garlits JE, et al. Hematopoietic cells and osteoblasts are derived from a common marrow progenitor after bone marrow transplantation. Proc Natl Acad Sci U S A. (2004); 101: 11761–11766. DOI: 10.1073/pnas.0404626101 38

Drosse I, Volkmer E, Capanna R., et al. Tissue Engineering for bone defect healing: An update and multi-component approach. Injury (2008); 39, S2: S9-S20. Review. DOI: 10.1016/S0020-1383(08)70011-1 51

Engler AJ, Rehfeldt F, Sen S, et al. "Micro-Tissue Elasticity: Measurements by Atomic Force Microscopy and its Influence on Cell Differentiation". In: DE Discher and Y-L Wang, Editors. Methods in Cell Biology: Cell Mechanics. (2007), Elsevier: New York. 521–545. 54

Engler AJ, Sen S, Sweeney HL, et al. Matrix elasticity directs stem cell lineage specification. Cell (2006); 126: 677. DOI: 10.1016/j.cell.2006.06.044 54

Faramawy AM. The Dynamic Role of Prosthetic Loading in Preserving Alveolar Ridges Restored by Tissue Engineering Biodegradable Polymer Implants Assisted by 3-D Computed Tomography. MSc thesis, Faculty of Dentistry, Alexandria University 2006. 54

Farrell E, Van der Jagt OP, Koevoet W, et al. Chondrogenic priming of human bone marrow stromal cells: a better route to bone repair? Tissue Eng Part C Methods. (2009); 15: 285–295. DOI: 10.1089/ten.tea.2008.0297 40

Forum S, and Orlowski W. Ridge preservation utilizing an alloplast prior to implant placement: Clinical and histological case reports. Pract. Periodontics Aesthet. Dent. 12, 393, 2000. 31

Fowler EB, Breault LG, and Rebitski G. Ridge preservation utilizing an acellular dermal allograft and demineralized freeze-dried bone allograft. Part II. Immediate edosseous implant placement. J. Periodontol. (2000); 71: 1360–4. DOI: 10.1902/jop.2000.71.8.1360 30

Fürst G, Gruber R, Tangl S, et al. Sinus grafting with autogenous platelet-rich plasma and bovine hydroxyapatite. A histomorphometric study in minipigs. Clin Oral Implants Res. (2003); 14(4): 500–8. 57

Gallego L, Junquera L, Garcia-Perez E, et al. Repair of rat mandibular bone defects by alveolar osteoblasts in a novel plasma-derived albumin scaffold. Tissue Eng Part A. 2010; 16(4): 1179-87. 36

Garg AK. Alveolar Ridge Preservation After Tooth Extraction. In: Bone Biology, Harvesting, Grafting for Dental Implants: Rational and Clinical Application. (2004) Quintessence Pub. pp 97–118. 29

Garrett RW and Emerson SG. Bone and blood vessels: the hard and the soft of hematopoietic stem cell niches. Cell Stem Cell. (2009); 4: 503–506. DOI: 10.1016/j.stem.2009.05.011 38, 39

Gerstner GE and Cianfarani T. Temporal dynamics of human masticatory sequences. Physiol Behav. (1998); 15, 64(4): 457–61. DOI: 10.1016/S0031-9384(98)00107-3 27

Gimbel M, Ashley RK, Sisodia M, et al. Repair of alveolar cleft defects: reduced morbidity with bone marrow stem cells in a resorbable matrix. J Craniofac Surg. (2007); 18(4): 895–901. 57

González-García R, Naval-Gías L and Rodríguez-Campo FJ. Distraction osteogenesis in the irradiated mandible for segmental mandibular reconstruction. J Oral Maxillofac Surg. (2009); 67(7): 1573–4. DOI: 10.1016/j.joms.2005.08.027 31

Haas R, Haidvogl D, Donath K, Watzek G. Freeze-dried homogeneous and heterogeneous bone for sinus augmentation in sheep. Part I: histological findings. Clin Oral Implants Res. (2002); 13(4):396–404. DOI: 10.1034/j.1600-0501.2002.130408.x 57

Hausmann E, Raisz LG, Miller WA. Endotoxin: stimulation of bone resorption in tissue culture. Science (1970); 168: 862–864. DOI: 10.1126/science.168.3933.862 29

Hibi H, Yamada Y, Ueda M, and Endo Y. Alveolar cleft osteoplasty using tissue-engineered osteogenic material. Int J Oral Maxillofac Surg. (2006); 35(6): 551–5. DOI: 10.1016/j.ijom.2005.12.007 57

Hildebolt CF. Osteoporosis and oral bone loss. Dentomaxillofac Radiol. (1997);26: 3–15. 29

Hoffmann O, Bartee BK, Beaumont C, et al. Alveolar bone preservation in extraction sockets using non-resorbable dPTFE membranes: a retrospective non-randomized study. J Periodontol. (2008); 79 (8): 1355–69. 52

Hollister SJ. Porous scaffold design for tissue engineering. Nat Mater. (2005); 4(7): 518–24. DOI: 10.1038/nmat1421 49

Horch HH, Pautke C. [Regeneration instead of reparation: a critical review of the autogenous bone transplant as "golden standard" of reconstructive oral surgery] Mund Kiefer Gesichtschir. (2006);10(4): 213–20. 30

Huang GT, Gronthos S, Shi S. Mesenchymal stem cells derived from dental tissues vs. those from other sources: their biology and role in regenerative medicine. J Dent Res. (2009); 88: 792–806. DOI: 10.1177/0022034509340867 36

Ilizarov GA and Ledyaev VI. The replacement of long tubular bone defects by lengthening distraction osteotomy of one of the fragments.(1969). Clin Orthop Relat Res. 1992; 280: 7–10. 31

Ingber DE, Mow VC, Butter D, et al. Tissue Engineering and developmental biology: Going biomimetic. Tissue Engineering. (2006); 12: 3265–3283. DOI: 10.1089/ten.2006.12.3265 32

Intini G. The use of platelet-rich plasma in bone reconstruction therapy. Biomaterials (2009), 30(28): 4956–4966. DOI: 10.1016/j.biomaterials.2009.05.055 45

Jahangiri L, Devlin H, Ting K, and Nishimura I. Current perspectives in residual remodeling and its clinical implications: A review. Journal of Prosthetic Dentistry. (1998); 80 (2): 224–237. 29

Jung RE, Pjetursson BE, Glauser R, et al. A systematic review of the 5-year survival and complication rates of implant-supported single crowns. Clin Oral Implants Res. 2008 Feb; 19(2): 119–30. Epub 2007 Dec 7. Review. DOI: 10.1111/j.1600-0501.2007.01453.x

Jung RE, Thoma DS, Hammerle CHF. Assessment of the potential of growth factors for localized alveolar ridge augmentation: a systematic review. (2007) *J Clin Periodontol.* 2008; 35 (Suppl. 8): 255–281. DOI: 10.1111/j.1600-051X.2008.01270.x 45, 56

Jung RE, Windisch SI, Eggenschwiler AM, et al. A randomized-controlled clinical trial evaluating clinical and radiological outcomes after 3 and 5 years of dental implants placed in bone regenerated by means of GBR techniques with or without the addition of BMP-2. Clin. Oral Impl. Res. 2009; 20(7): 660–666. DOI: 10.1111/j.1600-0501.2008.01648.x 45, 56

Kaigler D, Cirelli JA, Giannobile WV. Growth factor delivery for oral and periodontal tissue engineering. Expert Opin Drug Deliv. (2006); 3(5): 647–662. DOI: 10.1517/17425247.3.5.647 41

Kanczler JM, Oreffo ROC. Osteogenesis and angiogenesis: The potential for engineering bone. European Cells and Materials. (2008); 15: 100–114. 43

Kanyama M, Kuboki T, Akiyama K, Nawachi K, Miyauchi FM, Yatani H, Kubota S, Nakanishi T, Takigawa M. Connective tissue growth factor expressed in rat alveolar bone regeneration sites after tooth extraction. Archives of Oral Biology. (2003); 48: 723–730. DOI: 10.1016/S0003-9969(03)00153-5 33, 42, 43, 44

Kardong KV. Vertebrates: comparative anatomy, function, 4th ed. WCB/McGraw-Hill, Boston, MA. (2006): pp 500. 20

Kassolis JD, Rosen PS, Reynolds MA. Alveolar ridge and sinus augmentation utilizing platelet-rich plasma in combination with freeze-dried bone allograft: case series. J Periodontol. (2000); 71(10): 1654–61. DOI: 10.1902/jop.2000.71.10.1654 57

Kawanami M, Andreasen JO, Borum MK, Schou S, Hjorting-Hansen E, Kato H. Infraposition of ankylosed permanent maxillary incisors after replantation related to age and sex. Endod Dent Traumatol (1999): 15: 50–56. DOI: 10.1111/j.1600-9657.1999.tb00752.x 26

King GN, King N, Cruchley AT, Wozney JM, Hughes FJ. Recombinant Human Bone Morphogenetic Protein-2 Promotes Wound Healing in Rat Periodontal Fenestration Defects. J Dent Res. 2007; 76(8): 1460–1470. DOI: 10.1177/00220345970760080801 43

Kinoshita Y, Matsuo M, Todoki K, Ozono S, Fukuoka S, Tsuzuki H, Nakamura M, Tomihata K, Shimamoto T, Ikada Y. Alveolar bone regeneration using absorbable poly(L-lactide-co-epsilon-caprolactone)/beta-tricalcium phosphate membrane and gelatin sponge incorporating basic fibroblast growth factor. Int J Oral Maxillofac Surg. 2008; 37(3): 275–81. DOI: 10.1016/j.ijom.2007.11.010 44

Kirmeier R, Payer M, Wehrschuetz M, Jakse N, Platzer S, Lorenzoni M. Evaluation of three-dimensional changes after sinus floor augmentation with different grafting materials. Clin Oral Implants Res. (2008); 19(4): 366–72. DOI: 10.1111/j.1600-0501.2007.01487.x 57

Kishigami S, Mishina Y. BMP signaling and early embryonic patterning. Cytokine and Growth Factor Reviews. (2005); 16: 265–278. DOI: 10.1016/j.cytogfr.2005.04.002 41

Kitamura M, Nakashima K, Kowashi Y, Fujii T, Shimauchi H, Sasano T, Furuuchi T, Fukuda M, Noguchi T, Shibutani T, Iwayama Y, Takashiba S, Kurihara H, Ninomiya M, Kido J, Nagata T, Hamachi T, Maeda K, Hara Y, Izumi Y, Hirofuji T, Imai E, Omae M, Watanuki M, Murakami S. Periodontal tissue regeneration using fibroblast growth factor-2: randomized controlled phase II clinical trial. PLoS One. 2008; 3(7): e2611. DOI: 10.1371/journal.pone.0002611 44

Klemetti E. A review of residual ridge resorption and bone density. The Journal of Prosthetic Dentistry. (1996), 75; 5: 512–514. DOI: 10.1016/S0022-3913(96)90455-2 29

Knezovie-Zlatarie D, Celebic A, and Lazic B. Resorptive Changes of Maxillary and Mandibular Bone Structures in Removable Denture Wearers. Acta Stomat Croat (2002); 36: 261–265. 29

Korpi JT, Astrom P, Lehtonen N, Tjaderhane L, Kallio-Pulkkinen S, Siponen M, Sorsa T, Pirila E, Salo T. Healing of extraction sockets in collagenase-2 (matrix metalloproteinase-8)-deficient mice. Eur J Oral Sci. 2009; 117: 248–254. 32

Kozawa Y, Chisaka H, Iwasa Y, Yokota R, Suzuki K and Yamamoto H. Origin and evolution of cementum as tooth attachment complex. J oral biosci. (2005); 47(1), 25–32. DOI: 10.2330/joralbiosci.47.25

Kugimiya F, Kawaguchi H, Kamekura S, Chikuda H, Ohba Sh, Yano F, Ogata N, Katagiri T, Harada Y, Azuma Y, Nakamura K, and Chung U. Involvement of Endogenous Bone Morphogenetic Protein (BMP) 2 and BMP6 in Bone Formation. The journal of biological chemistry. 2005; 280(42): 35704–35712. DOI: 10.1074/jbc.M505166200 42

Kunz C, Adolphs N, Buescher P, Hammer B, Rahn B. Distraction osteogenesis of the canine mandible: the impact of acute callus manipulation on vascularization and early bone formation. J Oral Maxillofac Surg. (2005); 63(1): 93–102. DOI: 10.1016/j.joms.2004.07.008 31

Lalani Z, Wong M, Brey E M, et al. Spatial and temporal localization of FGF-2 and VEGF in healing tooth extraction sockets in a rabbit model. J. Oral maxillofa. Surg. 2005; 63: 1500–8. DOI: 10.1016/j.joms.2005.03.032 54

Lalani Z, Wong M, Brey EM, Mikos AG, Duke PJ. Spatial and temporal localization of transforming growth factor-beta1, bone morphogenetic protein-2, and platelet-derived growth factor-A in healing tooth extraction sockets in a rabbit model. (1061) *J Oral Maxillofac Surg.* 2003 Sep;61(9):1061–72. DOI: 10.1016/S0278-2391(03)00319-7 54

Lang NP, Aroujo M, Marring T. Alveolar bone formation. In: clinical periodontology and implant dentistry. 4th ed. 2003 pp 866–896. 32, 36

Langer R and Vacanti JP. Tissue Engineering. Science. (1993). 260: 920. 32

Lee J, Cuddihy MJ, and Kotov NA. Three-Dimensional Cell Culture Matrices: State of the Art. Tissue Engineering: Part B. (2008); 14: 61–86. DOI: 10.1089/teb.2007.0150 47, 48

Lee SY, Miwa M, Sakai Y, Kuroda, et al. In vitro multipotentiality and characterization of human unfractured traumatic hemarthrosis-derived progenitor cells: a potential cell source for tissue repair. J Cell Physiol. (2007); 3: 561–566. 38

Leknes KN, Yang J, Qahash M, Polimeni G, Susin C, Wikesjö M. Alveolar ridge augmentation using implants coated with recombinant human bone morphogenetic protein-7 (rhBMP-7/rhOP-1): radiographic observations. J Clin Periodontol. 2008a; 35(10): 914–9. 46

Leknes KN, Yang J, Qahash M, et al. Alveolar ridge augmentation using implants coated with recombinant human bone morphogenetic protein-2 rhBMP-2). Radiographic observations. Clin Oral Implants Res. 2008b; 19: 1027–33. DOI: 10.1111/j.1600-051X.2010.01554.x 46

Lekovic V, Camargo PM, Klokkevold PR, et al. preservation of alveolar bone in extraction sockets using bioabsorbable membrans. J periodontal (1998); 69: 1044–1049. 30, 51

Li H, Yan F, Lei L, et al. Application of autologous cryopreserved bone marrow mesenchymal stem cells for periodontal regeneration in dogs. Cells Tissues Organs. (2009); 190: 94–101. DOI: 10.1159/000166547 36

Li J, Li H, Shi L, et al. A mathematical model for simulating the bone remodeling process under mechanical stimulus. Dent Mater. (2007) Sep;23(9):1073–8. Epub 2006 Nov 29. DOI: 10.1016/j.dental.2006.10.004 56

Lin D, Li Q, Li W, et al. Mandibular bone remodeling induced by dental implant. J Biomech. (2010) Jan 19; 43(2): 287–93. Epub 2009 Oct 7. DOI: 10.1016/j.jbiomech.2009.08.024 56

Lin Sh, Roguin A, Metzger Z, et al. Vascular endothelial growth factor (VEGF) response to dental trauma: a preliminary study in rats. Dental Traumatology. 2008; 24: 435–438. DOI: 10.1111/j.1600-9657.2008.00608.x 43

Lindhe L, Karring T, and Araújo M. The Anatomy of Periodontal Tissues. In: Lindhe J, Lang NP, Karring T. clinical periodontology and implant dentistry. Fifth Edition. (2008); Blackwell Publishing co. p 56–58. 36

Liu Ch, Wu Z, Sun H-Ch. The Effect of Simvastatin on mRNA Expression of Transforming Growth Factor-β1, Bone Morphogenetic Protein-2 and Vascular Endothelial Growth Factor in Tooth Extraction Socket. Int J Oral Sci. 2009; 1(2): 90–98. 44, 54

Liu Y, Zheng Y, Ding G, et al. Periodontal ligament stem cell-mediated treatment for periodontitis in miniature swine. Stem Cells. (2008); 26: 1065–1073. DOI: 10.1634/stemcells.2007-0734 36

Lodish H, Berk A, Zipursky LS, et al. Molecular Cell Biology4th Edition. New York: W.H. Freeman and Company, 2002. 47

Lopez-Roldan A, Abad DS, Bertomeu IG, et al. Bone resorption processes in patients wearing overdentures: A 6-years retrospective study. Med Oral Patol Oral Cir Bucal. 2009; 14 (4): E203–9. 29, 51

Lu L, Yaszemski MJ, Mikos AG. TGF-beta1 release from biodegradable polymer microparticles: its effects on marrow stromal osteoblast function. J Bone Joint Surg Am 2001;83-A Suppl 1(Pt 2): S82–91. 48

Lu M and Rabie AB. Quantitative assessment of early healing of intramembranous and endochondral autogenous bone grafts using micro-computed tomography and Q-win image analyzer. Int J Oral Maxillofac Surg. (2004); 33: 369–376. DOI: 10.1016/j.ijom.2003.09.009 40

Lutolf MP, Hubbell JA. Synthetic biomaterials as instructive extracellular microenvironments for morphogenesis in tissue engineering. Nat Biotechnol. (2005); 23(1): 47–55. DOI: 10.1038/nbt1055 49

Lutz R, Park J, Felszeghy E, et al. Bone regeneration after topical BMP-2-gene delivery in circumferential peri-implant bone defects. Clin. Oral Impl. Res. 2008; 19: 590–599. DOI: 10.1111/j.1600-0501.2007.01526.x 45, 47

Maire M, Chaubet F, Mary P, et al. Bovine BMP osteoinductive potential enhanced by functionalized dextran-derived hydrogels. Biomaterials. (2005); 26(24): 5085–92. DOI: 10.1016/j.biomaterials.2005.01.020 50

Malkoç S, Iseri H, Karaman AI, et al. Effects of mandibular symphyseal distraction osteogenesis on mandibular structures. Am J Orthod Dentofacial Orthop. (2006); 130(5): 603–11. DOI: 10.1016/j.ajodo.2005.02.024 31

Malmgren B and Malmgren O. Rate of infraposition ofreimplanted ankylosed incisors related to age and growth in children and adolescents. Dent Traumatol (2002): 18: 28–36. 26

Malmgren B, Malmgren O, Andreasen JO. Alveolar bone development after decoronation of ankylosed teeth. Endodontic Topics (2006), 14, 35–40. DOI: 10.1111/j.1601-1546.2008.00225.x 27

Mao JJ, Giannobile WV, Helms JA, et al. Craniofacial tissue engineering by stem cells. J Dent Res. (2006); 85: 966–979. DOI: 10.1177/154405910608501101 40

Marco F, Milena F, Gianluca G, et al. Peri-implant osteogenesis in health and osteoporosis. Micron. (2005); 36(7–8): 630-44. DOI: 10.1016/j.micron.2005.07.008 54

Marei MK, Nouh SR, Fata MM, et al. Patent No. 23731 12/7/2003 – Academy of scientific research and Technology – Egypt. 54

Marei MK, Nouh SR, Saad MM, et al. Preservation and regeneration of alveolar bone by tissue-engineered implants. Tiss Eng. (2005); 11: 751–767. DOI: 10.1089/ten.2005.11.751 29, 36, 48, 49

Marei MK, Saad MM, El-Ashwah AM, et al. Experimental formation of periodontal structure around titanium implants utilizing bone marrow mesenchymal stem cells: a pilot study. J oral implantology (2009); 35(3): 106–129. 36, 54, 56

Marei MK, Nashaat DH, Sakr MR, et al. Alveolar bone mineral density as an indicator of skeletal osteoporosis in Pre- and Post-menopausal women. Tissue Engineering (2002); 8: 1246. DOI: 10.1563/1548-1336-35.3.106 26

Marei MK, Sakr MR, El-Backly RM, et al. Correlation between skeletal and mandibular bone mineral density in peri- and post- menopausal Egyptian females. Osteoporosis Int. (2003); 14 supplement 7 P S34. 26

Marks R. Philosophy and practice of autogenous bone grafting. Oral and maxillofacial surg clin north Am. (1993); 5: 599–612. 30

Marks Jr SC and Schroeder HE. Tooth eruption: theories and facts. Anat Rec (1996): 245: 374–393. DOI: 10.1002/(SICI)1097-0185(199606)245:2%3C374::AID-AR18%3E3.0.CO;2-M 26

Matsubara T, Suardita K, Ishii M, et al. Alveolar bone marrow as a cell source for regenerative medicine: differences between alveolar and iliac bone marrow stromal cells. J Bone Miner Res. (2005); 20: 399–409. DOI: 10.1359/JBMR.041117 40

Mauck RL, Byers BA, Yuan X, et al. Regulation of cartilaginous ECM gene transcription by chondrocytes and MSCs in 3/D culture in response to dynamic coading. Biomech Model Mechanobiol 2007; 6: 113. 54

McCarthy JG, Schreiber J, Karp N, et al. Lengthening the human mandible by gradual distraction. Plast Reconstr Surg. 1992; 89(1): 1–8; discussion 9–10. 31

McKee MD and Kaartinen MT. Regulation of Biomineralization by Bone Proteins and Their Assembly into Extracellular Matrices: Implications for Implant Osseointegration. In: Zarb G, Lekholm U, Albrektsson T and Tenenbaum H. Aging , Osteoporosis and Dental Implants. Quintessence Pub. Co, (2002); pp 191–206. 24

McMinn RMH, Hutchings RT. Head, Neck and Brain, A Color atlas of human anatomy, 2nd ed. Wolfe Publishing LTd, London WC1E7LT. 1991, pp:20. 22

Mendes RM, Silva GA, Lima MF, et al. Sodium hyaluronate accelerates the healing process in tooth sockets of rats. Arch Oral Biol. (2008); 53(12): 1155–62. DOI: 10.1016/j.archoralbio.2008.07.001 32, 44, 50, 53, 54

Merzel J, Novaes PD, Furlan S. The effects of local trauma to the enamel-related periodontal tissues in the eruption of the rat incisor. Archives of Oral Biology. (2000); 45: 323–333. DOI: 10.1016/S0003-9969(99)00136-3 22

Miiller E, Naharro M, Carlsson GE. what are the prevalence and incidence of tooth loss in adult and elderly population in Europe. Clinical oral implants res. (2007); 18: 2–14. DOI: 10.1111/j.1600-0501.2007.01459.x 51

Mizuno H, Hata K, Kojima K, et al. A novel approach to regenerating periodontal tissue by grafting autologous cultured periosteum. Tissue Eng. (2006); 12: 1227–1235. DOI: 10.1089/ten.2006.12.1227 36

Moriyama K, Sahara N, Kageyama T, et al. Scanning electron microscopy of the three different types of cementum in the molar teeth of the guinea pig Archives of Oral Biology. (2006); 51: 439–448. DOI: 10.1016/j.archoralbio.2005.07.001 22

Morrison SJ and Spradling AC. Stem cells and niches: mechanisms that promote stem cell maintenance throughout life. Cell. (2008); 132: 598–611. DOI: 10.1016/j.cell.2008.01.038 38

Morselli PG, Giuliani R, Pinto V, et al. Treatment of alveolar cleft performing a pyramidal pocket and an autologous bone grafting. J Craniofac Surg. (2009); 20: 1566–1570. DOI: 10.1097/SCS.0b013e3181b0dacd 39

Murray VK. Anterior ridge preservation and augmentation using a synthetic osseous replacement graft. Compend. Contin. Educ. Dent. (1998); 19: 69. 30

Nakajima K, Abe T, Tanaka M, et al. Periodontal tissue engineering by transplantation of multilayered sheets of phenotypically modified gingival fibroblasts. J Periodontal Res. (2008); 43(6): 681–688. DOI: 10.1111/j.1600-0765.2007.01072.x 39, 40, 43, 53

Nakashima M and Reddi AH. The application of bone morphogenetic proteins to dental tissue engineering. Nat Biotechnol. (2003); 21(9): 1025–32. DOI: 10.1038/nbt864 48, 50

Nanci A and Somerman MJ. Periodontium. In: Nanci N. Tencate's oral histology: development, structure, and function. 6th ed. Mosby, Inc. 2003. pp 240–274. 24

Nemcovsky CE and Serfaty V. Alveolar ridge preservationfollowing extraction of maxillary anterior teeth: Reporton 23 consecutive cases. J. Periodontol. 1996; 67: 390. 30

Nevins M, Mellonig JT. Enhancement of the damaged edentulous ridge to receive dental implants: a combination of allograft and the GORE-TEX membrane. Int J Periodontics Restorative Dent. (1992); 12(2): 96–111. 30

Nevins ML, Camelo M, Schupbach P, et al. Human histologic evaluation of mineralized collagen bone substitute and recombinant platelet-derived growth factor-BB to create bone for implant placement in extraction socket defects at 4 and 6 months: a case series. Int J Periodontics Restorative Dent. 2009; 29(2): 129–39. 43

Nicodemus GD and Bryant SJ. Cell encapsulation in biodegradable hydrogels for tissue engineering applications. Tissue Eng Part B Rev. (2008); 14(2): 149–65. DOI: 10.1089/ten.teb.2007.0332s 47

Norton MR, and Wilson J. Dental implants placed in extraction sites implanted with bioactive glass: Human histology and clinical outcome. Int. J. Oral Maxillofac. Implants (2002); 17: 249. 31

Ohya M, Yamada Y, Ozawa R, et al. Sinus floor elevation applied tissue-engineered bone. Comparative study between mesenchymal stem cells/platelet-rich plasma (PRP) and autogenous bone with PRP complexes in rabbits. Clin Oral Implants Res. 2005 Oct; 16(5): 622–9. DOI: 10.1111/j.1600-0501.2005.01136.x 57

Owen M and Friedenstein AJ. Stromal stem cells: marrow-derived osteogenic precursors. Ciba Found Symp. (1988); 136: 42–60. 36

Ozaki W, Buchman SR, Goldstein SA, et al. A comparative analysis of the microarchitecture of cortical membranous and cortical endochondral onlay bone grafts in the craniofacial skeleton. Plast Reconstr Surg. (1999); 104(1): 139–47. 30

Ozaki W, Buchman SR. Volume maintenance of onlay bone grafts in the craniofacial skeleton: micro-architecture versus embryologic origin. Plast Reconstr Surg. (1998); 102(2): 291–9. 30

Pieri F, Lucarelli E, Corinaldesi G, et al. Mesenchymal stem cells and platelet-rich plasma enhance bone formation in sinus grafting: a histomorphometric study in minipigs. J Clin Periodontol. (2008); 35(6):539–46. DOI: 10.1111/j.1600-051X.2008.01220.x 57

Ramakrishnana PR, Lin WL, Sodek J, et al. Synthesis of noncollagenous extracellular matrix proteins during development of mineralised nodules by rat periodontal ligament cells in vitro. Calcif Tissue Int. (1995); 57: 52–59. DOI: 10.1007/BF00298997 33

Ramseier CA, Abramson ZR, Jin Q, et al. Gene therapeutics for periodontal regenerative medicine. Dent Clin North Am. (2006) Apr; 50(2): 245–63. DOI: 10.1016/j.cden.2005.12.001 51

Rauch F, Lauzier D, Croteau S, et al. Temporal and Spatial Expression Of Bone Morphogenetic Protein-2, -4, And -7 During Distraction Osteogenesis In Rabbits. Bone. 2000; 26(6): 611–617. 42

Rawashdeh MA, Telfah H. Secondary alveolar bone grafting: the dilemma of donor site selection and morbidity. Br J Oral Maxillofac Surg. (2008); 46(8): 665–70. DOI: 10.1016/j.bjoms.2008.07.184 39, 57

Reddi AH. Morphogenesis and Tissue Engineering of Bone and Cartilage: Inductive Signals, Stem Cells, and Biomimetic Biomaterials. Tissue Engineering. (2000); 6(4): 351–359. DOI: 10.1089/107632700418074 40, 41, 42

Ripamonti U, Ramoshebi LN, Teare J, et al. The induction of endochondral bone formation by transforming growth factor-β3: experimental studies in the non-human primate Papio ursinus. J. Cell. Mol. Med. 2008; 12(3): 1029–1048. DOI: 10.3109/08977190009028971 44

Ripamonti U and Reddi AH. Morphogenetic Proteins in Craniofacial and Periodontal Bone Repair Growth and Morphogenetic Factors in Bone Induction: Role of Osteogenin and Related Bone Morphogenetic Proteins in Craniofacial and Periodontal Bone Repair. Crit Rev Oral Bio. Med. (1992); 3: 1–14. 41, 43

Ripamonti U, Ferretti C, Heliotis M. Soluble and insoluble signals and the induction of bone formation: molecular therapeutics recapitulating development. J Anat. (2006); 209: 447–468. DOI: 10.1111/j.1469-7580.2006.00635.x 40, 41

Ripamonti U, Petit J-C, Teare J. Cementogenesis and the induction of periodontal tissue regeneration by the osteogenic proteins of the transforming growth factor-b superfamily. J Periodont Res. (2009); 44: 141–152. DOI: 10.1111/j.1600-0765.2008.01158.x 44

Ripamonti U, Ma S, Reddi AH. The critical role of geometry of porous hydroxyapatite delivery system in induction of bone by osteogenin, a bone morphogenetic protein. Matrix. (1992); 12(3): 202–12. 48

Ritchey B and Orban B. The crests of the interdental alveolar septa. J Periodontology (1953); 24: 75–87 22, 24

Rivera-Hidalgo F. Wound Healing. In: Avery JK. Oral Development and Histology. (2001) 3rd edition. pp 390. 29

Robey PG and Bianco P. The use of adult stem cells in rebuilding the human face. J Am Dent Assoc. (2006); 137: S 961–972. 36

Rosen V. BMP and BMP Inhibitors in Bone. Ann. N.Y. Acad. Sci. (2006); 1068: 19–25. DOI: 10.1196/annals.1346.005 42

Sacchetti B, Funari A, Michienzi S, et al. Self-renewing osteoprogenitors in bone marrow sinusoids can organize a hematopoietic microenvironment. Cell. (2007); 131: 324–336. DOI: 10.1016/j.cell.2007.08.025 38, 39

Saffar JL, Lasfargues JJ, Cherruah M. Alveolar bone and alveolar process: the socket that is never stable. Periodontology (2000), 13: 76, 1997. DOI: 10.1111/j.1600-0757.1997.tb00096.x 26

Sammartino G, Tia M, Marenzi G, et al. Use of autogenous platelet-rich plasma (PRP) in periodontal treatment after extraction of impacted mandibular molars. J. Oral maxillofacial surg. 2005; 63: 766–770. DOI: 10.1016/j.joms.2005.02.010 54

Santos FA, Pochapski MT, Martins MC, et al. Comparison of Biomaterial Implants in the Dental Socket: Histological Analysis in Dogs. Clin Implant Dent Relat Res. (2010); 12(1): 18–25 DOI: 10.1111/j.1708-8208.2008.00126.x 48

Sawada Y, Hokugo A, Nishiura A, et al. A trial of alveolar cleft bone regeneration by controlled release of bone morphogenetic protein: an experimental study in rabbits. Oral Surg Oral Med Oral Pathol Oral Radiol Endod. (2009); 108(6): 812–20. DOI: 10.1016/j.tripleo.2009.06.040 46, 57

Scheller EL , Krebsbach PH and Kohn DH. Tissue Engineering: State of The Art in Oral Rehabilitation. Journal of Oral Rehabilitation. (2009); 36: 368–389. DOI: 10.1111/j.1365-2842.2009.01939.x 44, 45, 47, 48

Schieker M, Mulschler W. Bridging posttraumatic bony defects. Established and new methods. Unfallchirug (2006); 109: 715–732. DOI: 10.1007/s00113-006-1152-z 51

Schimming R, Schmelzeisen R. Tissue-engineered bone for maxillary sinus augmentation. J Oral Maxillofac Surg. (2004); 62(6): 724–9. DOI: 10.1016/j.joms.2004.01.009 60

Schmelzeisen R, Schimming R, Sittinger M. Making bone: implant insertion into tissue-engineered bone for maxillary sinus floor augmentation-a preliminary report. J Craniomaxillofac Surg. (2003); 31(1): 34–9. DOI: 10.1016/S1010-5182(02)00163-4 60

Schneider R. Prosthetic concerns about atrophic alveolar ridges. Postgrad. Dent. (1999); 6: 3. 30

Schofield R. The relationship between the spleen colony-forming cell and the haemopoietic stem cell. Blood Cells. (1978); 4: 7–25. 38

Schwartz Z, Goldstein M, Raviv E, et al. Clinical evaluation of demineralized bone allograft in a hyaluronic acid carrier for sinus lift augmentation in humans: a computed tomography and histomorphometric study. Clin Oral Implants Res. (2007); 18(2): 204–11. DOI: 10.1111/j.1600-0501.2006.01303.x 57

Schwartz-Arad D, Gulayev N, and Chaushu G. Immediate versus non-immediate implantation for full-arch fixed reconstruction following extraction of all residual teeth: A retrospective comparative study. J. periodontol. (2000); 400: 923. DOI: 10.1902/jop.2000.71.6.923 29

Schwarz F, Rothamel D, Herten M, et al. Lateral ridge augmentation using particulated or block bone substitutes biocoated with rhGDF-5 and rhBMP-2: an immunohistochemical study in dogs. Clin Oral Implants Res. 2008; 19(7): 642–52. DOI: 10.1111/j.1600-0501.2008.01537.x 44

Seeherman H, Wozney JM. Delivery of bone morphogenetic proteins for orthopedic tissue regeneration .Cytokine and Growth Factor Reviews. (2005); 16: 329–345. DOI: 10.1016/j.cytogfr.2005.05.001 46

Seeman E. Bone quality. Osteoporosis International. (2003); 14 (5), S3–S7. 26

Seo BM, Miura M, Gronthos S, et al. Investigation of multipotent postnatal stem cells from human periodontal ligament. Lancet. (2004); 364: 149–155. DOI: 10.1016/S0140-6736(04)16627-0 36

Serino G, Rao W, Iezzi G, et al. Polylactide and polyglycolide sponge used in human extraction sockets: bone formation following 3 months after its application. Clin Oral Implants Res. 2008; 19(1): 26–31. DOI: 10.1111/j.1600-0501.2007.01311.x 53

Shayesteh YS, Khojasteh A, Soleimani M, et al. Sinus augmentation using human mesenchymal stem cells loaded into a beta-tricalcium phosphate/hydroxyapatite scaffold. Oral Surg Oral Med Oral Pathol Oral Radiol Endod. (2008); 106(2): 203–9. DOI: 10.1016/j.tripleo.2007.12.001 61

Shin H, Jo S, Mikos AG. Biomimetic materials for tissue engineering. Biomaterials. (2003); 24(24): 4353–64. DOI: 10.1016/S0142-9612(03)00339-9 49

Simon BI, Zatcoff AL, Kong JJW, et al. Clinical and Histological Comparison of Extraction Socket Healing Following the Use of Autologous Platelet-Rich Fibrin Matrix (PRFM) to Ridge Preservation Procedures Employing Demineralized Freeze Dried Bone Allograft Material and Membrane. The Open Dentistry Journal. 2009; 3: 92–99. DOI: 10.2174/1874210600903010092 45, 54

Skalak R, Fox CF. Tissue engineering. Granlibakken, Lake Tahoe: Proc workshop; New York: Liss; (1988): 26–29. 32

Springer IN, Nocini PF, Schlegel KA, et al. Two techniques for the preparation of cell-scaffold constructs suitable for sinus augmentation: steps into clinical application. Tissue Eng. (2006); 12(9): 2649–56. DOI: 10.1089/ten.2006.12.2649 61

Stevens MM. Biomaterials for bone tissue engineering. Materials today. (2008); 11: 18–25. DOI: 10.1016/S1369-7021(08)70086-5 47

Storrie H, Mooney DJ. Sustained delivery of plasmid DNA from polymeric scaffolds for tissue engineering. Adv Drug Deliv Rev. (2006) 7; 58(4): 500–14. DOI: 10.1016/j.addr.2006.03.004 50

Sun XJ, Zhang ZY, Wang SY, et al. Maxillary sinus floor elevation using a tissue-engineered bone complex with OsteoBone and bMSCs in rabbits. Clin Oral Implants Res. (2008); 19(8): 804–13. DOI: 10.1111/j.1600-0501.2008.01577.x 61

Sweeney E, Campbell M, Watkins K, et al. Altered endochondral ossification in collagen X mouse models leads to impaired immune responses. (2008); 237: 2693–2704. DOI: 10.1002/dvdy.21594 40

Sy IP. Alveolar ridge preservation using a bioactive glass particulate graft in extraction site defects. Gen. Dent. (2002); 50: 66. 31

Tibesar RJ, Price DL and Moore EJ. Mandibular distraction osteogenesis to relieve Pierre Robin airway obstruction. Am J Otolaryngol. (2006); 27(6): 436–9. DOI: 10.1016/j.amjoto.2006.03.006 31

Ting K, Petropulos LA, Iwatsuki M, et al. Altered cartilage phenotype expressed during intramembranous bone formation. J Bone Miner Res. (1993); 8: 1377–1387. DOI: 10.1002/jbmr.5650081112 40

Tiwana PS, Kushner GM, Haug RH. Maxillary sinus augmentation. Dent Clin North Am. (2006); 50(3): 409–24. DOI: 10.1016/j.cden.2006.03.004 57

Tobita M, Uysal, AC, Ogawa R, et al. Periodontal tissue regeneration with adipose-derived stem cells. Tissue Eng. Part A. (2008); 14: 945–953. DOI: 10.1089/ten.2007.0048 36

Todo H. Healing mechanism of tooth extraction wounds in rats. I. Initial cellular response to tooth extraction in rats studied with 3 H-thymidine. Arch Oral Biol 1968; 13: 1421–1427. DOI: 10.1016/0003-9969(68)90024-1 29

Tortelli F, Tasso R, Loiacono F, et al. The development of tissue-engineered bone of different origin through endochondral and intramembranous ossification following the implantation of mesenchymal stem cells and osteoblasts in a murine model. Biomaterials. (2010); 31: 242–249. DOI: 10.1016/j.biomaterials.2009.09.038 39

Trombelli L, Farina R, Marzola A, et al. Modeling and remodeling of human extraction sockets. J Clin Periodontol. (2008); 35: 630–639. DOI: 10.1111/j.1600-051X.2008.01246.x 32, 33, 44

Tsumaki N and Yoshikawa H. The role of bone morphogenetic proteins in endochondral bone formation. *Cytokine Growth Factor Rev.* (2005); 16(3): 279–85. DOI: 10.1016/j.cytogfr.2005.04.001 41

Urist M. Bone: formation by autoinduction. Science. (1965); 150: 893–899. DOI: 10.1126/science.150.3698.893 41

Vogel RE and Wheeler SL. Tissue preservation for single tooth anterior esthetics. Compend. Contin. Educ. Dent. (2001); 22: 657. 30

Vunjak-Novakovic G and Kaplan DL. Tissue engineering the next generation. Tissue Engineering (2006), 12: 3261–3263. DOI: 10.1089/ten.2006.12.3261 32

Wang S, Zhang Z, Xia L, et al. Systematic evaluation of a tissue-engineered bone for maxillary sinus augmentation in large animal canine model. Bone. (2010); 46(1): 91–100. DOI: 10.1016/j.bone.2009.09.008 61

Weng Y, Wang M, Liu W, et al. Repair of experimental alveolar bone defects by tissue-engineered bone. Tiss Eng. (2006); 12: 1503–1513. DOI: 10.1089/ten.2006.12.1503 36, 54

Wheeler SL, Vogel RE, and Casellini R. Tissue preservation and maintenance of optimum esthetics: A clinical report. Int. J. Oral Maxillofac. Implants (2000); 15: 265. 30

Wiesen M. and Kitzis R. Preservation of the alveolar ridge at implant sites. Periodontal Clin. Invest. (1998); 20: 17. 30

Wikesjö UM, Qahash M, Polimeni G, et al. Alveolar ridge augmentation using implants coated with recombinant human bone morphogenetic protein-2: histologic observations. J Clin Periodontol. 2008; 35(11): 1001–10. DOI: 10.1111/j.1600-051X.2008.01321.x 46, 52

Wikesjö UME, Qahash M, Huang Y-H, et al. Bone morphogenetic proteins for periodontal and alveolar indications; biological observations – clinical implications. Orthod Craniofac Res. (2009); 12: 263–270. DOI: 10.1111/j.1601-6343.2009.01461.x 45

Wikesjö UME, Sorensen RG, Kinoshita A, et al. rhBMP-2 / a-BSM induces significant vertical alveolar ridge augmentation and dental implant osseointegration. Clin Implant Dent Relat Res. 2002; 4: 173–181. DOI: 10.1111/j.1708-8208.2002.tb00169.x 46

Williams R, Zager N. The periodontium. In: Shaw JH, Sweeney EA, Cappuccino CC and Meller S. Textbook of Oral Biology edition 1st. 1978; pp255–276. 23

Wood RM, Moore DL. Grafting of the maxillary sinus with intraorally harvested autogenous bone prior to implant placement. Int J Oral Maxillofac Implants. (1988); 3(3): 209–14. 57

Wu Z, Liu C, Zang G, et al. the effect of simvastatin on remodeling of the alveolar bone following tooth extraction. Int. J. Oral maxillofac. Surg. (2008); 37: 170–176. DOI: 10.1016/j.ijom.2007.06.018 54

Wyatt CC. The Effect of Prosthodontic Treatment on Alveolar Bone Loss: a Review of Literature. J prosthet dent (1998); 80: 362–366. DOI: 10.1016/S0022-3913(98)70138-6 29

Yamada Y, Nakamura S, Ito K, et al. Injectable tissue-engineered bone using autogenous bone marrow-derived stromal cells for maxillary sinus augmentation: clinical application report from a 2–6-year follow-up. Tissue Eng Part A. 2008 Oct; 14(10): 1699–707. DOI: 10.1089/ten.tea.2007.0189 56, 57

Yamasaki K, Miura F, Suda T. Prostaglandin as a mediator of bone resorption induced by experimental tooth movement in rats. J Dent Res. (1980), 59: 1635–1642. DOI: 10.1177/00220345800590101301 30

Yang J, Lee HM, Vemino A. Ridge preservation of dentition with sever periodontitis. Compend contin educ. Dent. (2000); 21: 579–583 30, 54

Yang Y, El Haj AJ. Biodegradable scaffolds..delivery systems for cell therapies. Expert Opin Biol Ther. (2006); 6(5): 485–98. DOI: 10.1517/14712598.6.5.485 47

Yao J, Li X, Bao C, et al. A novel technique to reconstruct a boxlike bone defect in the mandible and support dental implants with In vivo tissue-engineered bone. J Biomed Mater Res B Appl Biomater. (2009) Nov; 91(2): 805–12. DOI: 10.1002/jbm.b.31459 56

Yen SL, Yamashita DD, Gross J, et al. Combining orthodontic tooth movement with distraction osteogenesis to close cleft spaces and improve maxillary arch form in cleft lip and palate patients. Am J Orthod Dentofacial Orthop. 2005; 127(2): 224–32. DOI: 10.1016/j.ajodo.2003.09.036 57

Yonezawa H, Harada K, Ikebe T, et al. Effect of recombinant human bone morphogenetic protein-2 (rhBMP-2) on bone consolidation on distraction osteogenesis: a preliminary study in rabbit mandibles. J Craniomaxillofac Surg. 2006; 34(5): 270–6. DOI: 10.1016/j.jcms.2006.02.003 31

Younger EM and Chapman MW. Morbidity at bone graft donor sites. J orthop trauma (1989); 3: 192–195. 30

Zaky SH, Cancedda R. Engineering Craniofacial Structures: Facing the Challenge. J Dent Res. (2009); 88(12): 1077–1091. DOI: 10.1177/0022034509349926 36, 40, 45

Zhang Y, Song J, Shi B, et al. Combination of scaffold and adenovirus vectors expressing bone morphogenetic protein-7 for alveolar bone regeneration at dental implant defects. Biomaterials. 2007; 28(31): 4635–4642. DOI: 10.1016/j.biomaterials.2007.07.009 45

Zhang Y, Wang Y, Shi B, et al. A platelet-derived growth factor releasing chitosan/coral composite scaffold for periodontal tissue engineering. Biomaterials. (2007); 28(8): 1515–22. DOI: 10.1016/j.biomaterials.2006.11.040 50

Zhang Z, Song Y, Zhang X, et al. Msx1/Bmp4 genetic pathway regulates mammalian alveolar bone formation via induction of Dlx5 and Cbfa1. Mechanisms of Development. 2003; 120: 1469–1479. DOI: 10.1016/j.mod.2003.09.002 20, 41, 51

Zizelmann C, Schoen R, Metzger MC, et al. Bone formation after sinus augmentation with engineered bone. Clin Oral Implants Res. (2007); 18(1): 69–73. 60

CHAPTER 3

Tissue Engineering of the Periodontal Tissues

Ugo Ripamonti, MD, PhD

Jean-Claude Petit, BSc, LDS, MDent

June Teare, MSc

3.1 CHAPTER SUMMARY

The mechanistic understanding of the molecular and cellular pathways of cell differentiation resulting in tissue induction and morphogenesis is central to tissue engineering and morphogenesis. The induction of bone formation, after the classic studies of '*Bone: Formation by autoinduction*' (Urist, 1965; Reddi and Huggins, 1972), has set into motion the ripple-like cascade of tissue induction and morphogenesis including the induction of the complex tissue morphologies of the periodontal tissues, i.e., the avascular cementum attached to the dentinal root surface, the alveolar bone and the periodontal ligament fibers uniting the alveolar bone to the root cementum penetrating the cementum as mineralized Sharpey's fibers. The induction of bone formation and the induction of periodontal tissue regeneration require three key components: a soluble osteogenic molecular signal, an insoluble signal or substratum that delivers the osteogenic signal and acts as a scaffold for the induction of bone formation, and responding stem cells capable of differentiation into selected cellular phenotypes after transmembrane serine-threonine kinase receptors' activation and phosphorilation ultimately leading to tissue induction and morphogenesis. The osteogenic proteins of the transforming growth factor-β (TGF-β) supergene family, the bone morphogenetic/osteogenic proteins (BMPs/OPs) uniquely in the non-human primate *Papio ursinus*, the three mammalian TGF-β isoforms initiate the induction of endochondral bone formation when implanted in heterotopic extraskeletal sites. Periodontal tissue regeneration is induced by invocation of the soluble osteogenic molecular signals of the TGF-β supergene family, which, when combined with insoluble signals or substrata, trigger the ripple-like cascade of bone and periodontal ligament induction, cementogenesis with the faithful insertion of Sharpey's fibers into the newly formed cementum tightly anchoring into the mineralized dentinal matrix. Soluble osteogenic molecular signals combined with different insoluble signals or substrata induce periodontal tissue regeneration in a variety of pre-clinical animal models including non-human primates. The translational research learned in pre-clinical animal models still has to show beyond doubt that the soluble osteogenic molecular signals of the TGF-β supergene family also engineer cementogenesis with functionally oriented periodontal ligament fibers in human patients. The greatest challenge of all will be the selection of the recombinant morphogens following the discovery of the pleiotropic activity together with the apparent redundancy of molecular signals initiating the induction of bone formation in non-human primate species. Of the now available recombinant human proteins, which are the specific isoforms that will induce predictable periodontal tissue regeneration in humans? Does the presence of multiple molecularly different but homologous isoforms have a therapeutic significance? Molecularly different but homologous isoforms implanted in non-human primates *Papio ursinus* have shown a structure/activity profile that result in the induction of different tissue morphologies in the periodontal context; this has indicated that the apparent redundancy of osteogenic soluble molecular signals orchestrates the final tuning of the vast pleiotropic activity of the osteogenic proteins acting as body morphogenetic proteins (BMPs) (Reddi, 2005).

3.2   INTRODUCTION: THE INDUCTION OF BONE FORMATION

Millions of years of evolution have set the emergence of *Homo sapiens*' cranio-mandibulo-facial skeleton, its masticatory apparatus with a full complement of 32 permanent teeth tightly locked into the alveolar bone by a periodontal ligament system which unites the alveolar bone to the root cementum in an unique gomphosis exquisitely operated by the periodontal tissues, the gingivae, the alveolar bone, the periodontal ligament system, and the root cementum.

The dentition and its supporting periodontal apparatus has been Nature's complex evolutionary challenge from the seemingly simple elasmobranchial's gomphosis (Fig. 3.1) to the complex functionalities and molecular/anatomical pathways of the periodontal tissues from the extinct Australopithecinae and Homo species to extant primates including *Homo sapiens*. Besides, similar if not equal periodontal structures, early hominids also suffered from attachment loss, pocket formation, and chronic periodontitis including early onset prepubertal periodontitis as seen on the fossilized gnathic remains of the bipedal Australopithecinae, including a juvenile specimen of *Australopithecus africanus* affected by a suggested case of prepubertal periodontitis (Fig. 3.2) (Ripamonti, 1988, 1989; Ripamonti and Petit, 1991).

Molars, premolars, canines and incisors are locked into the mandibular and maxillary alveolar bone by a superior design of architecture and tissue engineering. Cross-linked periodontal ligament fibers insert and penetrate the root cementum as well as the underlying highly mineralized dentine extracellular matrix as bona fide Sharpey's fibers, thus uniting the masticatory dental elements to the alveolar bone (Bhaskar SN, 1980), providing the mechanical support during mastication, deglutition, copulation as well as many other physiological functions of the human life, not lastly, the extraordinary power of the human smile.

A cardinal rule of contemporary tissue engineering and regenerative medicine is pattern formation and the attainment of tissue form and function or morphogenesis (Reddi, 1984). A

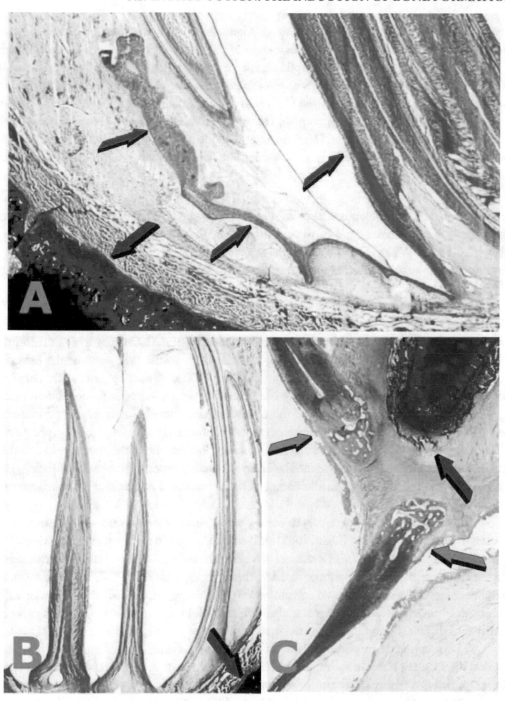

Figure 3.1: *Caption on the next page.*

Figure 3.1: *Caption for figure on previous page.* Continuous tooth eruption and morphogenesis in adolescent sharks *Charcharinus obscurus* jaws caught off the Indian Ocean North Coast of South Africa. *Magenta* arrows point to the dental eruption lamina mechanistically providing the continuously erupting teeth as shown in A (right) and B, moving mesially to the end of the cartilaginous structure (*royal blue* arrows in A, B, and C) of the shark jaw. C: Continuously migrating dental elements reach the end of the cartilage strut of the jaw (*royal blue* arrow) and the attachment apparatus is lost, to a number of teeth; exfoliation and loss of attachment (*blue* arrows in C) of the most mesial, i.e., anterior teeth, provide space distally for the continuously erupting shark dentition. Undecalcified sections cut at 7μm stained with toluidine blue.

central question in developmental biology and thus tissue engineering and regenerative medicine at large is the molecular basis of pattern formation, tissue induction and morphogenesis (Reddi, 1984, 2000a,b). What is the molecular basis of morphology? The cellular and molecular basis of morphology is tissue induction and morphogenesis, i.e., the genesis of form and function (Reddi, 1984, 2000a).

Classic studies on 'Bone: Formation by autoinduction' (Urist, 1965) have set the critical rules for tissue engineering and regenerative medicine; major breakthroughs in periodontal tissue regeneration have come from the molecular dissection of the fascinating phenomenon of "Bone: Formation by autoinduction" (Urist, 1965; Reddi and Huggins, 1972) and of the bone induction principle (Urist et al., 1967, 1968; Urist and Strates, 1971). Which are the molecular signals that initiate the ripple-like cascade of tissue induction, cementogenesis, and periodontal ligament regeneration with the final assembly of the complex tissue morphologies of the periodontal tissues restoring the severed dento-alveolar junction? Morphogenesis is induced by morphogens (Turing, 1952); as a prerequisite to tissue induction and morphogenesis, there must exist several signaling molecules, or morphogens, first defined by Turing as form-generating substances (Turing, 1952), that initiate tissue induction and the attainment of tissue form and function or morphogenesis (Turing, 1952; Reddi, 1984, 2000a,b).

The aim of this monograph is to convey a concise perspective on the phenomenon of 'Bone: Formation by autoinduction' (Urist, 1965) and its applications in the context of periodontal tissue engineering (Ripamonti, 2007). The rational of linking the induction of periodontal tissue regeneration to the induction of bone formation is based on the discovery that tissue induction and morphogenesis require three key components: an osteoinductive soluble signal; an insoluble signal or substratum, which delivers the osteogenic signal and acts as a scaffold for new bone formation; and host cells capable of differentiating into the osteoblastic phenotype (Reddi, 2000b; Ripamonti et al., 2000a).

The emergence in post-natal life of the complex tissue morphologies of the mammalian periodontal tissues rests on a simple and fascinating concept: morphogens exploited in embryonic development are re-exploited and re-deployed for the initiation of postnatal tissue induction and morphogenesis (Levander, 1938; Reddi, 1984, 2000b; Ripamonti, 2003, 2007; Ripamonti et al., 2000a, 2006a). Which are the soluble molecular signals that initiate the ripple-like cascade of tissue

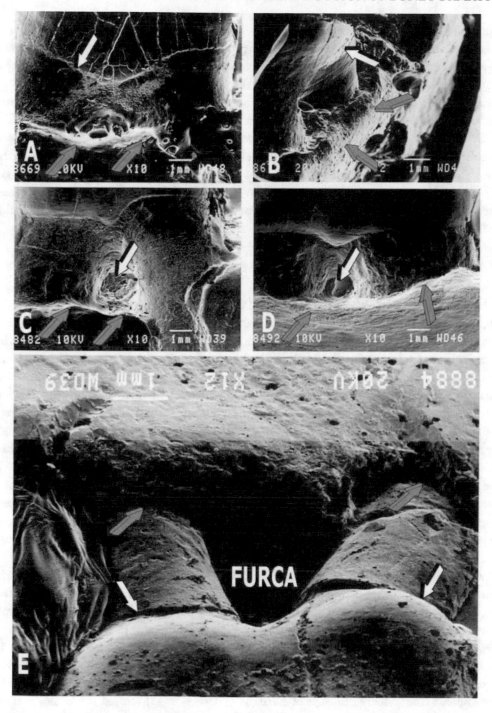

Figure 3.2: *Caption on the next page.*

Figure 3.2: *Caption for figure on previous page.* The hard evidence of alveolar bone loss in fossilized gnathic remains at the Pleio-Pleistocene boundary of *Australopithecus robustus* (A, B) and *Australopithecus africanus* (C, D) including a juvenile A. *africanus* specimen (E) affected by a suggested case of prepubertal periodontitis 2 to 3 million-years before the present (Ripamonti, 1988). A, B: Scanning electron microscopy photomacrographs of alveolar bone loss with furcation exposure from the cemento-enamel junction (white arrows) to the remaining alveolar bone (*blue* arrows) of mandibular molars of A. *robustus*. C, D: Significant furcation exposure (white arrows) in A. *africanus* adult specimens with resorption of the alveolar bony housings (*blue* arrows). E: Prominent horizontal bone loss completely exposing the furca of the 1st deciduous maxillary molar of A. *africanus* Sts 24a juvenile specimen affected by a suggested case of prepbertal periodontitis (Ripamonti, 1988). White arrows in E indicate the cemento-enamel junction and *blue* arrows the remaining alveolar bony housings highlighting the extent of horizontal bone loss suffered by the juvenile gracile Australopithecus (Ripamonti, 1988).

induction and regeneration of the periodontal tissues in rodents, non-human and human primates? Or, how is the induction of bone formation initiated?

Last century, research has witnessed an important scientific and technological drive to isolate, purify to homogeneity and to clone the human proteins responsible for the too long, elusive 'osteogenetic activity of a variety of mineralized and non-mineralized extracellular matrices.' Matrices include bone, dentine, the uroepithelium, the urinary bladder, and the muscular wall of the ileum (Ripamonti et al., 2006a) as well as the kidney after ligation of the renal artery as reported by the classic studies of Sacerdotti and Frattin (1901). In his important studies, Gustav Levander stated that when devitalized acidified alcohol/extracted bone is transplanted in heterotopic sites of recipient animals, a 'specific bone substance is liberated from the bone and is carried by the tissue lymph to the surrounding areas where it is able to activate the mesenchymal tissue in such a way that this becomes differentiated into bone tissue – either directly or by means of the embryonic pre-existing stage of bone and cartilaginous tissue' (Levander, 1938). Levander subsequently reported that a 'substance extracted by alcohol from the skeletal tissue has the power to activate the non-specific mesenchymal tissue into the formation of bone tissue' (Levander, 1938; Willestaedt et al., 1950). Levander significantly contributed to the further understanding of the bone induction principle by introducing the term 'induction' stating that 'the circumstance that a tissue is able to affect another in a specifically differentiating direction I have termed 'induction' – a term borrowed from embryology, introduced, as is well known, by Spemann and his school at the turn of the century' (Levander, 1945). The osteogenic potency of various intact or partially extracted matrices became known as 'osteogenic activity' (Urist, 1965; Urist et al., 1967, 1968; Reddi and Huggins, 1972) previously defined by Levander and his school as 'specific bone forming substance' (Levander, 1938). Following experiments using alcohol extracts of cartilage, (Lacroix, 1945) termed the 'osteogenic activity' as 'osteogenin' whilst Moss 'osteogenic inductor' with osteogenic activity in the brain (Moss, 1958).

Collectively, the important work of several different scientists across the globe including Sacerdotti and Frattin (1901), Willestaedt et al. (1950), Levander (1938, 1945), Lacroix (1945) and Moss (1958) strongly indicated that the chemical nature of the inductor(s) was proteic in nature and paved the way to the systematic studies on the 'bone induction principle' (Urist et al., 1967, 1968; Urist and Strates, 1971) leading to the introduction of a hypothetical bone morphogenetic protein complex within the bone matrix ultimately responsible for the bone induction cascade (Urist and Strates, 1971).

Several landmark experiments of the last century have thus hypothized that unknown substances named 'inductors', 'osteogenins' or 'bone morphogenetic proteins' (Sacerdotti and Frattin, 1901; Levander, 1938, 1945; Lacroix, 1945; Willestaedt et al., 1950; Moss, 1958; Urist and Strates, 1971; Ripamonti, 2003; Ripamonti et al., 2006a) were embedded within the extracellular matrices of several mineralized and non-mineralized tissues and organs including bone, dentine, uroepithelium and the kidney and could be released by extraction and/or demineralization of the implanted matrix (Urist, 1965; Levander, 1938). Importantly, the acid test for the induction of bone formation was and still is the implantation of putative osteogenic proteins in extraskeletal heterotopic sites to prove beyond doubt whatsoever that the endochondral cascade is initiated by the treated bone matrices and/or protein preparations when implanted in extraskeletal sites (Urist, 1965).

Which are the molecular signals that initiate the *de novo* induction of heterotopic bone formation? The classic studies of Levander (1938, 1945); Willestaedt et al. (1950); Lacroix (1945); Moss (1958); Urist (1965), and Urist et al. (1967, 1968), have indicated that the inductors, either named osteogenins or bone morphogenetic proteins, are tightly bound to the extracellular matrix of several mineralized and non-mineralized matrices of disparate *and diverse* organs and tissues as diverse as bone, dentine, uroepithelium, brain, heart and kidney. How to identify and isolate the inductors, the putative and elusive osteogenic proteins found within such a different and complex variety of extracellular matrices?

3.3   DISSOCIATIVE EXTRACTION AND RECONSTITUTION OF THE BONE MATRIX COMPONENTS: RESTORATION OF THE ENDOCHONDRAL BONE INDUCTION CASCADE

A fundamental step forward was the introduction of the dissociative extraction and reconstitution of the extracellular matrix components of the bone matrix (Sampath and Reddi, 1981, 1983). The intact demineralized bone matrix is biologically active in both the heterotopic intramuscular and the subcutaneous space of rodents yielding the multistep cascade of chemotaxis, cellular recruitment and differentiation, mitosis and cellular transformation into the chondro-osteoblastic phenotype (Urist, 1965; Reddi and Huggins, 1972; Reddi, 1984, 2000a,b). The intact demineralized bone matrix has been dissociatively extracted into an insoluble signal, mainly collagenous matrix, and into soluble protein fractions containing the putative osteogenic proteins responsible for the induction of bone formation in heterotopic extraskeletal sites (Sampath and Reddi, 1981, 1983).

The bone matrix is in the solid state (Reddi, 1997); the mineralized extracellular matrix of bone, mainly collagenous, contains minute amounts of non-collagenous glycoproteins tightly bound to both the organic and inorganic matrices of bone (Reddi, 1997). The observation that the bone matrix in the solid state is insoluble (Reddi, 1997) has been a stumbling block for the identification of putative osteogenic proteins within the bone matrix. '*The matrix in the solid state*' has been finally resolved by the dissociative extraction and reconstitution of the extracellular matrix components (Sampath and Reddi, 1981). The chaotropic extraction of the bone matrix yielded an insoluble inactive mainly collagenous matrix residue and a glycoprotein-rich extract from the bone matrix (Sampath and Reddi, 1981); implantation of either the insoluble or the soluble signals in the subcutaneous space of the rat did not result in the induction of endochondral bone formation (Sampath and Reddi, 1981) and the biological activity of the intact demineralized bone matrix is abolished (Sampath and Reddi, 1981). Restoration of the biological activity lost after the dissociative extraction in chaotropic agents, could be achieved by reconstituting the insoluble and soluble signals thus restoring the endochondral bone induction cascade of the intact demineralized bone matrix (Sampath and Reddi, 1981).

No single experiment has been so critical for the rapid advance of '*the bone induction principle*' (Urist et al., 1967) as initiated by intact demineralized bone matrix, now dissociatively extracted and reconstituted with the available matrix components (Sampath and Reddi, 1981, 1983). The next step was the understanding that the putative osteogenic proteins dissociatively extracted from the extracellular matrix of bone were homologous between mammalian species and that the alloantigenic load was essentially contained within the respective insoluble collagenous matrices when reconstituted with allogeneic and/or xenogeneic soluble signals (Sampath and Reddi, 1983). The experiments of Sampath and Reddi (1983) were instrumental for the understanding that the elusive but now finally identified osteogenic proteins in the bone matrix were homologous across mammalian species (Sampath and Reddi, 1983). The osteogenic soluble signals could be implanted across mammalian species provided, however, that the reconstitution was performed with the allogeneic insoluble signal of the animal species to be eventually implanted (Sampath and Reddi, 1983). The dissociative extraction and reconstitution of the bone matrix components solubilized the putative osteogenic protein fractions, which were then subjected to adsorption/affinity and molecular sieve gel filtration chromatography (Luyten et al., 1989; Ripamonti et al., 1992, 1993; Ripamonti, 2006). The dissociative extraction and reconstitution of the matrix components also provided a *bona fide* bioassay for testing protein fractions endowed with osteoinductive activity (Sampath and Reddi, 1983); the bioassay was instrumental for the final purification to homogeneity of osteogenic protein fractions of naturally-derived bone morphogenetic/osteogenic proteins (BMPs/OPs) (Ripamonti, 2006); BMPs/OPs were isolated and purified in sufficient quantities and purity to provide amino acid sequence information (Ripamonti, 2006). Full length complementary DNA clones were then isolated, encoding the human equivalent of several BMPs/OPs (Wozney et al., 1988; Özkaynak et al., 1990; Sampath et al., 1992; Ripamonti, 2006).

Significantly, the BMPs/OPs are members of a family of proteins that have potent and diverse effects on cell proliferation and differentiation, body axis formation, growth control, and sexual development in many organisms, the transforming growth factor-β (TGF-β) superfamily (Wozney et al., 1988; Özkaynak et al., 1990; Sampath et al., 1992; Ripamonti et al., 2005; Ripamonti, 2006). The dissociative extraction and reconstitution of the bone matrix components (Sampath and Reddi, 1981) was also instrumental for the realization that a matrix carrier is required to deliver the osteogenic activity of the several osteogenic proteins of the TGF-β superfamily (Ripamonti, 2003) when implanted in heterotopic extraskeletal and orthotopic intraskeletal defect sites in different animal models where the induction of bone formation is required (Wozney et al., 1988; Özkaynak et al., 1990; Sampath et al., 1992; Ripamonti et al., 2000a, 2005, 2006a; Ripamonti, 2003, 2006).

The use of xenogeneic BMPs/OPs yet delivered by allogeneic insoluble signals has been deployed to demonstrate the pleiotropic activity of highly purified naturally-derived BMPs/OPs which, when combined with allogeneic insoluble collagenous signals induce periodontal tissue regeneration in non-human primates of the species *Papio ursinus* (Ripamonti et al., 1994, 2009a; Ripamonti and Reddi, 1997; Ripamonti, 2007), as well as the induction of bone formation in rodents (Luyten et al., 1989). The operational reconstitution also induces heterotopic endochondral bone formation and the induction of bone formation in calvarial defects of the non-human primate *Papio ursinus* when allogeneic collagenous matrix is reconstituted with μg amounts of *Papio ursinus* highly purified naturally-derived osteogenin (Ripamonti et al., 1992, 1993).

3.4   NATURALLY-DERIVED BONE MORPHOGENETIC PROTEINS INDUCE CEMENTOGENESIS AND THE INDUCTION OF PERIODONTAL LIGAMENT REGENERATION

The osteogenic proteins of the TGF-β superfamily (Ripamonti, 2003) regulate tooth morphogenesis at different stages of development (Vainio et al., 1993; Hogan, 1996; Åberg et al., 1997; Thesleff and Sharpe, 1997; Thomadakis et al., 1999). The induction of cementogenesis, periodontal ligament and alveolar bone differentiation are regulated by the coordinated expression of BMPs/OPs and TGF-β family members (Vainio et al., 1993; Hogan, 1996; Åberg et al., 1997; Thesleff and Sharpe, 1997; Thomadakis et al., 1999). Nature relies on common yet limited molecular mechanisms to sustain the emergence of specialized tissues and organs (Reddi, 2000b; Ripamonti, 2003, 2007). Tissue induction in postnatal life recapitulates events, which occur in the normal course of embryonic development (Ripamonti, 2007). Immunolocalization of BMPs/OPs during tooth morphogenesis indicates that the secreted gene products play morphogenetic roles during cementogenesis and the assembly of a functionally oriented periodontal ligament system (Thomadakis et al., 1999; Ripamonti, 2007).

The expression patterns of several BMPs/OPs during tooth morphogenesis shows that there is co-distribution between specific family members; the localization of BMP-2, BMP-3 and osteogenic protein-1 (OP-1) also known as BMP-7 (Özkaynak et al., 1990; Åberg et al., 1997) during tooth morphogenesis shows the pleiotropic activity of BMPs/OPs beyond bone namely the developmental stages of mantle dentin deposition, cementogenesis and the induction of periodontal ligament formation (Thomadakis et al., 1999). We have learned that members of selected superfamily of proteins regulate bone induction and morphogenesis in embryonic development (Reddi, 2000b; Ripamonti, 2007); we have later learned that tissue induction and morphogenesis in postnatal life recapitulate embryonic development (Reddi, 2000b; Levander, 1938); we have learned thus that morphogenetic messages, i.e., the soluble osteogenic molecular signals of the TGF-β superfamily (Reddi, 2000a; Ripamonti, 2003), when deployed in post-natal life, initiate the induction of tissue morphogenesis as a recapitulation of embryonic development (Reddi, 2000a; Ripamonti, 2003, 2007).

The extraordinary induction of cementogenesis, periodontal ligament and alveolar bone regeneration by highly purified naturally-derived BMPs/OPs purified greater than 70,000 fold with respect to the crude guanidinium extracts of bovine matrix, are morphogenetic messages recapitulating embryonic development during tooth morphogenesis (Fig. 3.3) (Reddi, 2000a; Ripamonti et al., 1994, 2005; Ripamonti, 2007). Highly purified naturally-derived BMPs/OPs isolated from bovine bone matrices in conjunction with allogeneic insoluble collagenous bone matrices as carrier induce periodontal ligament and alveolar bone regeneration with inserting Sharpey's fibers into the newly formed and mineralized cementum and dentine (Figs. 3.3 A,B,C,D,F).

Undecalcified sections cut at 3μm show the newly deposited cementoid matrix as collagenic material as yet to be mineralized along the exposed dentine matrix (Figs. 3.3 D,E). Foci of mineralization are seen scattered throughout the cementoid matrix with the emergence of Sharpey's fibers protruding from the cementoid matrix as yet to be mineralized (Fig. 3.3). Sharpey's fibers are essential constituents of the periodontal ligament system uniting the dental element to the alveolar bone by inserting into the newly formed cementum as well as anchoring as mineralized fibers into the dentine extracellular matrix (Fig. 3.3).

We have also shown by using high power digital microphotography of the periodontal ligament fibers (Ripamonti et al., 2009a) that the fibers within the periodontal ligament space are used for locomotion of progenitor cells resident within the perivascular/vascular *niche* of the periodontal ligament space (Ripamonti et al., 2009a); progenitor cells thus ride individual fibers and locomote to the alveolar bone and/or to the cementum side of the periodontal ligament space (Ripamonti et al., 2009a). Final differentiation with the induction of the cementoblastic and/or fibroblastic phenotype on one side of the periodontal ligament space or the induction of the osteoblastic phenotype on the opposite side is regulated by morphogens gradients across the periodontal ligament system (Ripamonti et al., 1994; Ripamonti, 2007; Ripamonti et al., 2009a). Which are the stem cell progenitors in the periodontal ligament complex that attach to individual principal fibers riding the fiber to the cementum or to the bony side of the periodontal ligament space?

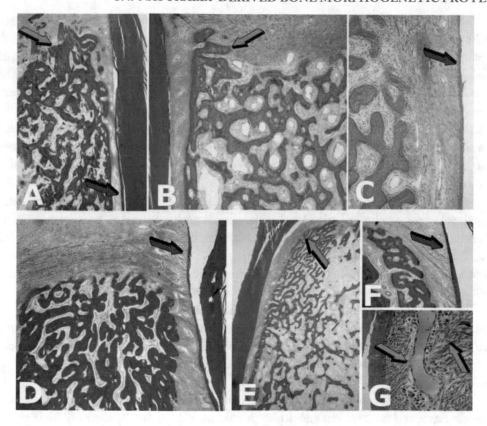

Figure 3.3: Periodontal tissue induction and morphogenesis by highly purified naturally-derived osteogenic fractions purified greater than 70,000 fold with respect to crude guanidinium extracts of bovine bone matrices [27, 35]. A, B: substantial induction of mineralized alveolar bone (*blue* arrows) surfaced by osteoid seams. Induction of cementogenesis (*magenta* arrows) against the surgically instrumented root surface. C: Detail of mineralized bone in blue surfaced by osteoid seams populated by contiguous osteoblasts; *magenta* arrow points to cementoid matrix as yet to mineralized populated by secreting cementoblasts. D, E: Prominent osteogenesis with almost complete regeneration of the alveolar bone in Class II furcation defects surgically exposed in *Papio ursinus*; *magenta* arrow in D indicated newly formed mineralized cementum; *blue* arrow in E points to osteoid seams surfacing newly formed mineralized bone in *blue*. F: Detail of tissue induction and morphogenesis of mineralized cementum (*magenta* arrow), Sharpey's fibers and trabeculae of newly formed and mineralized bone in blue surfaced by osteoid seams. G: Angiogenesis within the newly formed periodontal ligament with Sharpey's fibers inserting into newly formed cementum; *blue* arrows point to the blending of Sharpey's fibres into the basement membrane of the capillary as well as periodontal ligament fibers that provide the supramolecular tridimensional assembly for mesenchymal stem cells riding individual fibers from the vascular compartment to the alveolar bone and/or to the cemental sides of the periodontal ligament complex. Undecalcified sections cut at 3 to 4μm stained free-floating with a modified Goldner's trichrome.

Highly purified naturally-derived BMPs/OPs when implanted into furcation defects prepared in the mandibular molars of *Papio ursinus* directly induce significant angiogenesis and capillary invasion within the newly regenerated periodontal ligament space (Figs. 3.4 A,B,C) (Ripamonti, 2006; Ripamonti et al., 2006a, 2009a). There is no tissue induction without angiogenesis; high power views of capillary invasion within the periodontal ligament space show that periodontal ligament fibers and Sharpey's fibers blend into the extracellular matrix of the basement membrane of the invading capillaries (Figs. 3.4 D,E). The interaction of the periodontal ligament fibers with the invading capillaries provide the anatomical structure for continuously migrating progenitor cells riding the fibers from the capillary/angiogenic compartment to both sides of the periodontal ligament space (Fig. 3.4).

The operational reconstitution of the solubilized osteogenic molecular signals with an insoluble signal or substratum pelletized molecular signals onto substrata to be implanted in a variety of heterotopic and orthotopic sites providing a bioassay for the identification of *bona fide* initiators of bone differentiation including periodontal tissue regeneration (Ripamonti, 2007; Ripamonti et al., 2009a). The induction of tissue formation or morphogenesis is the essence of the tissue engineering paradigm engineered by combinatorial molecular protocols whereby soluble molecular signals are combined and reconstituted with insoluble signals or substrata (Ripamonti, 2007; Ripamonti et al., 2009a); the reconstitution or recombination of the osteogenic soluble molecular signals with different insoluble signals or substrata when implanted in heterotopic and orthotopic sites of animal models triggers the ripple-like cascade of bone differentiation by induction (Reddi, 2000b; Ripamonti, 2003).

Undecalcified sections cut at 3μ of Class II furcation defects of mandibular molars prepared in adult *Papio ursinus* have indicated how highly purified naturally-derived BMPs/OPs are also inducers of angiogenesis predating the bone induction cascade (Figs. 3.3, 3.4) (Ripamonti, 2007; Ripamonti et al., 2009a). Furcation defect specimens show prominent angiogenesis as a prerequisite to osteogenesis (Figs. 3.4 I,J,L,M). Tissue induction initiates as mesenchymal cell condensations surrounding capillaries which have invaded the exposed furcation defect after the implantation of $250\mu g$ of highly purified bovine BMPs/OPs (Ripamonti et al., 1994, 2009a; Ripamonti, 2007). The invading capillaries are organizers of the osteonic cortico/cancellous structure of the bone unit or osteosome (Reddi, 1997) by providing progenitor mesenchymal cells to differentiate into osteoblastic-like cells. Newly formed trabeculae of mesenchymal condensations, yet to be mineralized, are concentrically engineered around single blood vessels which ultimately dictate the pattern of tissue induction and morphogenesis. Further remodeling of the mesenchymal cellular condensations results in focal mineralization with the morphogenesis and patterning of the osteonic-like bone with osteoid seams populated by contiguous osteoblasts (Figs. 3.4 I,J,L,M) (Ripamonti, 2006, 2007; Ripamonti et al., 2009a); osteoblasts are in close relationship with the '*organogenic*' capillaries with patterning function during organogenesis firstly described by Aristotle (Crivellato et al., 2007) and later defined '*angiogenic*' and '*osteogenic*' by the unique classic morphological studies of Trueta et al. (1963).

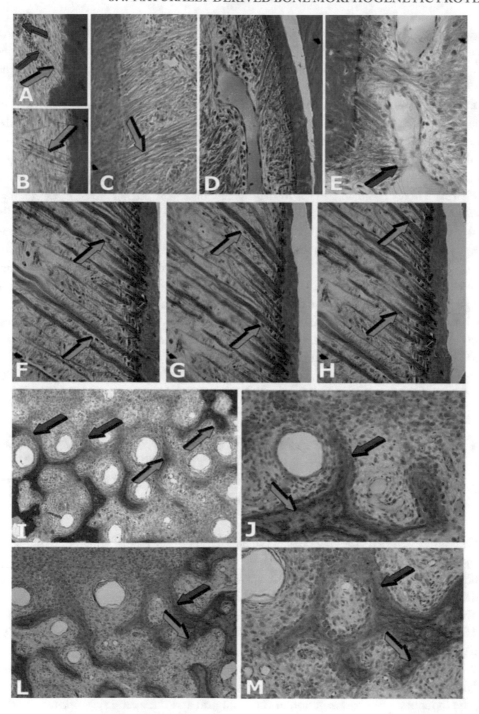

Figure 3.4: *Caption on the next page.*

Figure 3.4: *Caption for figure on previous page.* Naturally-derived highly purified bone morphogenetic/osteogenic proteins (OPs/OPs): Tissue induction and morphogenesis of the periodontal ligament fibres, angiogenesis, capillary sprouting and cementogenesis with vascular endothelial/perycitic stem cells riding individual Sharpey's fibres inserting into newly formed and mineralized cementum. A, B: Induction of Sharpey's fibres (*blue* arrows) inserting into mineralized dentine surfaced by secreting cementoblasts; A: mesenchymal condensations (*magenta* arrows) of stem/precursor cells with condensed chromatin predating angiogenesis and capillary sprouting within the newly formed periodontal ligament complex. C: Organized periodontal ligament space with Sharpey's fibres inserting into mineralized dentine; D: angiogenesis and capillary sprouting within the newly formed periodontal ligament space; E: Sharpey's fibres blend and attach to the capillary basement membrane of the sprouting capillaries almost touching the large hyperchromatic endothelial cells of the osteogenetic vessels of Truetas morphologic analyses (Trueta et al., 1963). F, G, H: High power view panels of Sharpey's fibres inserting into newly formed mineralized cementum against the instrumented dentinal surface; *blue* arrows indicate riding cells that ride individual Sharpey's fibres from the vascular endothelial/perycitic compartment of the periodontal ligament to the cementoid surface of the periodontal ligament space. I, J, K, L: Invasion of *osteogenetic vessels* as induced by implanted naturally-derived highly purified BMPs/OPs induce cellular condensations around each invading capillary; mesenchymal condensations surround each invading capillary (*magenta* arrows J) and differentiate contiguous osteoblastic-like cells facing the central blood vessels providing the supramolecular assembly of the induction of the primate cortico/cancellous osteonic bone. Foci of mineralization (*blue* arrows) construct the beginning of the primate osteonic bone. Undecalcified sections cut 2 to 3μm stained free-floating with modified Goldner's trichrome.

3.5 REDUNDANCY AND STRUCTURE/ACTIVITY PROFILE OF THE OSTEOGENIC PROTEINS OF THE TRANSFORMING GROWTH FACTOR-β SUPERGENE FAMILY

More than 15 years ago, the challenge of the biological significance of apparent redundancy has been formulated (Ripamonti and Reddi, 1994); as predicted, the choice of a suitable protein amongst redundant initiators of bone differentiation is still a formidable challenge to periodontologists and skeletal recostructionists alike (Ripamonti and Reddi, 1994, 1997; Ripamonti, 2007). More importantly perhaps, the presence of multiple molecularly different but homologous BMPs/OPs expressed and localized within several tissues, including the developing periodontium, reflects different function *in vivo* and have a therapeutic significance (Ripamonti and Reddi, 1994, 1997; Ripamonti, 2007; Ripamonti et al., 2009a). To date, more than 40 related proteins with BMP/OP-like sequences and activities have been sequenced and cloned; little is known, however, about their interactions during the induction of bone formation or the biological and therapeutic significance of this apparent redundancy (Ripamonti and Reddi, 1994; Ripamonti, 2006, 2007; Ripamonti et al., 2009a).

BMPs/OPs are involved in inductive events that control pattern formation during morphogenesis in such disparate tissues and organs as the brain, the teeth, with the associated cementum and periodontal ligament system, the eye, the lungs, the heart, the kidney, the pancreas, the liver and the skin as pleiotropic soluble signals during organogenesis, maintenance and regenerative phenomena (Vainio et al., 1993; Hogan, 1996; Åberg et al., 1997; Thesleff and Sharpe, 1997; Thomadakis et al., 1999). These strikingly multiple pleiotropic effects, deserving the terminology of body morphogenetic proteins (BMPs) (Reddi, 2005), spring from minor amino acid sequence variations in the carboxy-terminal domain of the proteins (Ripamonti et al., 2000a, 2004, 2006b; Ripamonti, 2003) as well as in the transduction of distinct signal pathways by individual Smad proteins after serine-threonine kinase receptor activation and expression of the down-steam regulators Smad-6 and -7 (Ripamonti, 2003; Ripamonti et al., 2004, 2006a,b).

Amino acid sequence variations in the active carboxy-terminal domain of each protein confer specialized pleiotropic activities to each BMP/OP isoform; the above is the molecular basis that determines the structure/activity profile of each molecular isoforms.

The biological significance of redundancy and its therapeutic implication rests on developing a structure/activity profile of each single recombinant isoform, i.e., to *study in vivo* the morphogenetic drive of each single and structurally different recombinant BMPs/OPs and in primates only so as to discover pleiotropic activities of different protein isoforms based on specific amino acid sequences (Fig. 3.5) (Ripamonti and Reddi, 1994; Ripamonti, 2003; Ripamonti et al., 2006a,b).

Members of the TGF-β and BMP/OP families have a synchronous but temporally and spatially different expression and localization during periodontal tissue morphogenesis (Thomadakis et al., 1999); this has indicated a mosaic pattern of gene product's expression during periodontal tissue morphogenesis (Vainio et al., 1993; Hogan, 1996; Åberg et al., 1997; Thesleff and Sharpe, 1997; Thomadakis et al., 1999). The synchronous but temporarily and spatially different expression and localization of TGF-β super family members during tooth and periodontal morphogenesis point to novel therapeutic approaches using the now available recombinant human proteins, since post-natal tissue induction and morphogenesis recapitulate embryonic development (Levander, 1938; Reddi, 2000b; Ripamonti, 2003).

The finding of synchronous but spatially different BMPs/OPs during periodontal tissue morphogenesis as indeed suggested a novel therapeutic approach using morphogen combinations based on recapitulation of embryonic development (Ripamonti et al., 2004). Twelve furcation defects prepared in the 1st and 2nd mandibular molars of three non-human primate *Papio ursinus* were used to assess whether quantitative histological aspects of periodontal tissue regeneration could be enhanced and tissue morphogenesis modified by combined or single applications of recombinant hOP-1 and hBMP-2 (Ripamonti et al., 2001). As expected from previous studies (Ripamonti et al., 1996) 60 days after healing, hOP-1/treated defects showed substantial cementogenesis with scattered collagenous matrix as carrier (Fig. 3.6 B) (Ripamonti et al., 2001). As previously reported (Ripamonti et al., 1996), there was a presence of a pseudo-ligament space between the newly formed mineralized matrix within the implanted furcation defect and the remaining alveolar bony housings (Figs. 3.5 H,I,J)

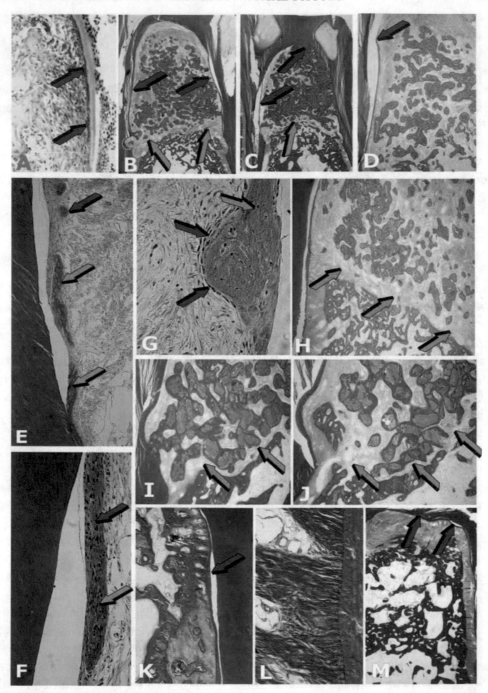

Figure 3.5: *Caption on the next page.*

Figure 3.5: *Caption for figure on previous page.* Induction of substantial cementogenesis by human recombinant osteogenic protein-1 (hOP-1) on surgically-exposed (B-J) and periodontally-induced (K-M) furcation defects in the adult non-human primate *P. ursinus*. A: Immunolocalization of OP-1 in the developing embryonic molar of a mouse pup prominently localized during cementogenesis (*magenta* arrows) and the genesis of Sharpey's fibres along the newly deposited dentine. B, C, D: Cementogenesis on surgically created furcation defects (*magenta* arrows) treated with 0.1 and 0.5 mg hOP-1 60 days after implantation; top arrow (C) points to the furca with mineralized cementoid material attached to the exposed dentine surface surrounding insoluble collagenous bone matrix as carrier; *Blue* arrows point the pseudo-ligament space between the remaining alveolar bone and the newly formed cemental matrix but not bone within the implanted furcation surrounding scattered remnants of the collagenic matrix as carrier. E, F, G: prominent cementogenesis as condensed cellular collagenic material along the exposed dentinal surface; areas of as yet to be mineralized newly deposited cementoid matrix (*magenta* arrows) are interspersed with foci of mineralization (*blue* arrows) along the newly formed cementum. H, I, J: details of the pseudo-ligament space (*blue* arrows) between the remaining alveolar bony housing (dark blue apical) and the newly formed and mineralized cemental matrix within the treated furcation defects; K, L, M: Tissue induction and morphogenesis with *rerstitutio ad integrum* of the periodontal tissues after long term implantation of 0.5 (K, L) and 2.5 mg gamma-irradiated hOP-1 per 500 mg of bovine insoluble collagenous bone matrix as carrier in periodontally-induced furcation defects in *Papio ursinus* (Ripamonti et al., 2002). Continuous cementogenesis up to the furca (*magenta* arrows in M) six months after implantation of the recombinant osteogenic device. Undecalcified sections cut 5 to 7μm stained free-floating with modified Goldner's trichrome.

(Fig. 3.6 B) indicating that the newly formed mineralized matrix surrounding scattered remnants of the collagenous matrix as carrier is essentially newly formed cementum, thus not uniting with the alveolar bone proper (Figs. 3.5 H,I,J) (Fig. 3.6 B) (Ripamonti et al., 1996, 2001).

hBMP-2, applied singly, induced greater amounts of mineralized bone and osteoid when compared to single doses of hOP-1 (Figs. 3.6 E,F) (Ripamonti et al., 2001). Binary applications of the recombinant proteins induced greater amounts of cementogenesis on the exposed root surfaces (Fig. 3.6 C) (Ripamonti et al., 2001). The results of the study, which is the first published report to address *in vivo* and in primate models the structure/activity profile of hBMP/OP family members, has shown that tissue morphogenesis induced by hOP-1 and hBMP-2 is qualitatively different when morphogens are applied singly or in binary application (Ripamonti et al., 1996,?, 2001, 2004, 2006b). When implanted in either surgically-created (Ripamonti et al., 1996,?, 2001) or periodontally induced (Ripamonti et al., 2002) furcation defects of *Papio ursinus*, hOP-1 is highly cementogenic. Long-term studies after implantation of 0.5 and 2.5 mg hOP-1 in periodontally-induced furcation defects of *Papio ursinus* showed induction of periodontal tissue regeneration with extensive cementogenesis with insertion of Sharpey's fibres (Figs. 3.5 L,M,N) (Ripamonti et al., 2002).

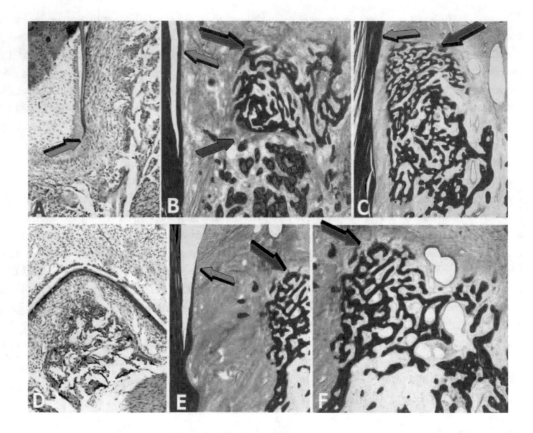

Figure 3.6: Pleiotropic activity and structure/activity profile of recombinant human osteogenic protein-1 (hOP-1) (A,B), recombinant human bone morphogenetic protein-2 (hBMP-2) (D,E, F) and binary application of hOP-1 and BMP-2 (C) harvested 60 days after implantation in surgically-created Class II furcations defects of mandibular molars of *Papio ursinus*. A: Expression of OP-1 in mantle dentine (*blue* arrow) and during the induction of cementogenesis along the newly deposited dentine. B: Substantial cementogenesis (*blue* arrow) along the exposed dentine; induction of newly formed bone surfaced by osteoid seams but induction of mineralized cementum-like tissue not in continuity with the newly formed bone (*royal blue* arrow). D: Immunolocalization of BMP-2 in erupting mouse pup molar highlighting the localization of the protein essentially to the alveolar bone; E, F: Tissue induction and morphogenesis by single application of hBMP-2; induction of newly formed mineralized bone (in blue) surfaced by osteoid seams (*magenta* arrows) with limited induction of cementogenesis (*blue* arrow in E); C: restoration of the induction of cementogenesis (*blue* arrow) and alveolar bone regeneration with osteoid seams surfacing newly formed mineralized bone (*magenta* arrow) by binary applications of 100μg hOP-1 and 100μg hBMP-2. Undecalcified sections cut 2 to 3μm stained free-floating with modified Goldner's trichrome.

In contrast, hBMP-2 is highly osteogenic when implanted in periodontal defects of a variety of animal models including rodents, canines and non-human primates (Ripamonti, 2007; Ripamonti et al., 2009a). In Class II furcation defects of *Papio ursinus*, hBMP-2 is preferentially osteogenic with limited cementogenesis (Ripamonti et al., 2001). The binary application of hOP-1 and hBMP-2 restores cementogenesis with the induction of prominent osteoid seams populated by contiguous osteoblasts surfacing newly formed mineralized bone (Fig. 3.6 C) (Ripamonti et al., 2001).

In a canine study, histometric analyses at 5 months after implantation of doses of recombinant hBMP-2 showed that cementum regeneration was less than control treatment without hBMP-2 (Choi et al., 2002); this and other studies have indicated that the recombinant hBMP-2 does not have a significant effect on cementum regeneration and formation of a functional oriented periodontal ligament system (Sigurdsson et al., 1995a,b; Giannobile et al., 1998; Choi et al., 2002). hBMP-2 inhibits differentiation and mineralization of cementoblasts *in vitro* (Zhao et al., 2003); exposure of cementoblasts to hBMP-2 *in vitro* results in a dose-dependent reduction of bone sialoprotein and collagen I gene expression, and inhibition of cell-induced mineral nodule formation (Zhao et al., 2003).

3.6 CHALLENGES: STEM CELLS, RIDING CELLS AND THE SYNERGISTIC INDUCTION OF BONE FORMATION

The induction of bone formation together with the induction of predictable periodontal tissue regeneration needs now to be based on the expression patterns of morphogenetic gene products as a recapitulation of embryonic development (Zhao et al., 2003; Ripamonti, 2007; Ripamonti et al., 2009a). The mechanistic understanding of molecular signals in solution interacting with insoluble signals of the extracellular matrix has provided the conceptual framework for tissue induction and morphogenesis as combinatorial protocols of soluble osteogenic molecular signals with insoluble biomimetic signals activated by responding stem cells (Reddi, 2000b; Ripamonti, 2003, 2007; Ripamonti et al., 2004, 2009a). Combinatorial molecular protocols with insoluble signals or substrata acting on responding stem cells are the essence of the tissue engineering paradigm (Ripamonti and Reddi, 1994; Reddi, 2000b; Ripamonti, 2003, 2007; Ripamonti et al., 2004, 2007, 2009a).

Comprehensive research data have explored the presence of multipotent postnatal stem cells in the human periodontal ligament (Melcher et al., 1987; Seo et al., 2004; Bartold et al., 2000; Lin et al., 2008; Ripamonti and Petit, 2009). Stem cells are critical for the induction of periodontal tissue regeneration with a direct involvement in periodontal tissue engineering (Seo et al., 2004; Bartold et al., 2000; Lin et al., 2008). Ultimately, the addition of stem cells to periodontal defects will help to enhance the induction of cementogenesis with *de novo* Sharpey's fibres inserted into newly formed cementum, the essence of periodontal tissues induction (Ripamonti, 2007; Ripamonti and Petit, 2009). The classic studies of Melcher et al. (1987) have suggested that cementoblasts, osteoblasts, and their progenitors that are found in the periodontal ligament space have their origin from the endosteal spaces of the alveolar process (Melcher et al., 1987). Expression of different phenotypes and generation of cementum or alveolar bone may be dependent on whether a common lineage of progenitor cells attach on residual cemental or exposed dentinal substrata or stay in the alveolar bone side of the periodontal ligament space (Melcher et al., 1987; Ripamonti et al., 1994; Ripamonti, 2007; Ripamonti and Petit, 2009). The presence of exposed dentine may preferentially modulate the expression of the cementogenic phenotype on readily available cementoblasts and their progenitors attached to dentine or residing within the periodontal ligament space (Melcher et al., 1987; Ripamonti, 2007; Ripamonti et al., 2009a).

In a series of systematic studies in the different microenvironments of the non-human primate *Papio ursinus*, including the calvarial, mandibular, periodontal and the *rectus abdominis* sites, we have learned that primate tissues and microenvironments respond remarkable differently when compared to rodents and lagomorphs (Ripamonti et al., 2009b). Intramuscular heterotopic implantation of the mammalian TGF-β isoforms induce substantial bone formation by induction in the non-human primate *Papio ursinus* (Ripamonti, 2003; Ripamonti et al., 1997, 2000b, 2008, 2004). Implantation of TGF-β_3 in surgically created furcation defects induces cementogenesis and alveolar bone re-generation (Teare et al., 2008; Ripamonti et al., 2009a). The addition of morcellized fragments of autogenous *rectus abdominis* muscle with available myoblastic stem cells significantly enhance the induction of bone formation by the hTGF-β_3 isoform (Ripamonti et al., 2009b) and cementogenesis in Class III furcation defects implanted with doses of hTGF-β_3 in Matrigel matrix (Ripamonti et al., 2009c).

Further anatomical and molecular studies should be devoted to understand the induction of angiogenesis and its intimate association with the periodontal ligament fibers; the capillary and peri-capillary space provide a *niche* of progenitor stem cells from the vascular compartment; per-icytes and/or endothelial cells migrate from the vascular compartment into the periodontal liga-ment space by riding periodontal ligament fibers towards the cementogenic and/or osteogenic sides of the periodontal ligament space ultimately differentiating by morphogen gradients across the ride (Ripamonti et al., 2009a). The '*ride*' of progenitor stem cells across morphogen gradients at-tached to single periodontal ligament fibers is the basic supramolecular assembly for the maintenance and regeneration of the periodontal ligament space.

Tissue engineering has made it possible to tackle the interdisciplinary challenge to *de novo* construct new tissues and organs so as to regenerate lost parts of the human body (Reddi, 2000b; Ripamonti et al., 2004). This great challenge has been investigated by using the basic set of rules that regulate the '*bone induction principle*' (Urist et al., 1967; Reddi, 2000b); the induction of bone formation has provided substantial cues for the identification of novel therapeutic approaches to induce cementogenesis and alveolar bone regeneration with the faithful insertion of Sharpey's fibers into newly formed cementum (Ripamonti et al., 2004); ultimately, however, tissue engineers and periodontologists alike should now devise novel biomimetic matrices that *per se* are able to set into motion the expression and synthesis of selected gene products of the TGF-β superfamily as secreted by the very patients when implanted with bioactive biomimetic matrices that *per se* differentiate responding stem cells into osteoblastic/cementobalstic cell lines secreting bone and cemental matrix to achieve self-induced and self-controlled periodontal tissue regeneration, even without the exogenous application of the soluble molecular signals of the TGF-β supergene family (Ripamonti, 2004).

3.7   ACKNOWLEDGMENTS

The Bone Research Laboratory work on '*Bone: Formation by autoinduction*' has been constantly supported by the Medical Research Council of South Africa, the University of the Witwatersrand, Johannesburg, and the National Research Foundation since its inception at the medical school of the university. I acknowledge *ad hoc* grants and human recombinant morphogens to the Bone Research Laboratory after sponsored or non-sponsored research projects with Creative BioMolecules-Striker Biotech USA, Novartis AG Switzerland, Genzyme Corporation, USA. A U.S. PHS National Institutes of Health Foreign Grant no. DE 107 12-01 jointly with Johns Hopkins University, partly supported craniofacial bone regeneration in primates using recombinant human osteogenic proteins. I would like to acknowledge with thanks the special work of Barbara van den Heever for preparing impeccable undecalcified sections of the dentin/cementum and periodontal ligament/alveolar bone complex which took the Bone Research Laboratory into another dimension of the fascinating phenomenon of '*Bone: Formation by autoinduction*.' The attached digital images of undecalcified sections of bone/dentine/cementum complex testify of the uniqueness of Barbara's work. Importantly, Barbara has had the opportunity to teach her unique techniques and skills to senior technologists in the Bone Research Laboratory: J. Crooks, J. Teare, L. Renton and R. Parak. Remembering with fondness research studies many years ago at the National Institute of Dental Research, NIH Bethesda, Maryland, I would like to thank Laura Yates for the purification of cartilaginous extracts of shark chondrocrania and vertebral bodies in the search of putative osteogenic proteins in cartilaginous fishes and Jean-Claude Petit for the help of implanting recombinant hOP-1 and naturally-derived osteogenin intramuscularly in *Charcharinus obscurus* sharks at the Oceanographic Research Institute, Durban, and the Natal Shark Board for providing fast boats to fish the adolescent sharks. I would like to acknowledge with thanks the mentorship and scientific drive of Drs Marshall Urist and Hari A. Reddi for sharing their passion, knowledge and expertise on '*bone: formation by autoinduction*.'

BIBLIOGRAPHY

Åberg, T., J. Wozney, T.Thesleff. (1997) *Expression patterns of bone morphogenetic proteins (BMPs) in the developing mouse tooth suggest roles in morphogenesis and cell differentiation.* Dev Dyn, 1997. **210**(4): p. 383–396.
DOI: 10.1002/(SICI)1097-0177(199712)210:4%3C383::AID-AJA3%3E3.0.CO;2-C   91, 92, 97

Bartold P.M., S. Shi, and S. Gronthos, (2000) *Stem cells and periodontal regeneration.* Periodontology 2000, 2006. **40**: p. 164–172. DOI: 10.1111/j.1600-0757.2005.00139.x 101

Bhaskar S.N., editor. Orban's Oral Histology and Embryology; CV Mosby Co., London, (1980). p. 204–239. 84

Choi S-H., C-K. Kim, K-S. Cho, J.S. Huh, R.G. Sorenson, and J.M. Wozney et al, *Effect of recombinant human bone morphogenetic protein-2 absorbable collagen sponge (rhBMP-2/ACS) on healing in 3-wall intrabony defects in dogs.* J Periodontol, 2002. **73**(1): p. 63–72. DOI: 10.1902/jop.2002.73.1.63 101

Crivellato E., B. Nico, and D. Ribatti, (2007) *Contribution of endothelial cells to organogenesis : a modern reappraisal of an old Aristotelian concept.* J Anat, 2007. **211**(4): p. 415–427. DOI: 10.1111/j.1469-7580.2007.00790.x 94

Giannobile W.V., S. Ryan, M.S. Shih, D.L. Su, P.L. Kaplan, and T.C. Chan, (1998) *Recombinant human osteogenic protein-1 (OP-1) stimulates periodontal wound healing in class III furcation defects.* J Periodontol, 1998. **69**(2): p. 129–137. 101

Hogan B.L.M., (1996) *Bone morphogenetic proteins; multifunctional regulators of vertebrate development.* Genes Dev, 1996. **10**(13): p. 1580–1594. DOI: 10.1101/gad.10.13.1580 91, 97

Lacroix P., (1945) *Recent investigations on the growth of bone.* Nature, 1945. **156**: p. 576. DOI: 10.1038/156576a0 88, 89

Levander G., (1938) *A study of bone regeneration.* Surg Gynecol and Obstet, 1938. 67: p. 705–714. 86, 88, 89, 92, 97

Levander G., (1945) *Tissue induction.* Nature, 1945. **155**: p. 148–149. DOI: 10.1038/155148a0 88, 89

Lin N-H., D. Menicanin, K. Mrozik, S. Gronthos, and P.M. Bartold, (2008) *Putative stem cells in regenerating human periodontium.* J Periodont Res, 2008. **43**(5): p. 514–523. DOI: 10.1111/j.1600-0765.2007.01061.x 101

Luyten F.P., N.S. Cunningham, S. Ma, N. Muthukumaran, R.G. Hammonds, and W.B. Nevins, et al., (1989) *Purification and partial amino acid sequence of osteogenin, a protein initiating bone differentiation.* J Biol Chem, 1989. **264**(23): p. 13377–13380. 90, 91

Melcher A.H., C.A. McCulloch, T. Cheong, E. Nemeth, and A. Shiga, (1987) *Cells from bone synthesize cementum-like and bone-like tissue in vitro and may migrate into periodontal ligament in vivo.* J Periodontal Res, 1987. **22**: p. 246–247. DOI: 10.1111/j.1600-0765.1987.tb01579.x 101, 102

Moss M.L., (1958) *Extraction of an osteogenic inductor factor from bone.* Science, 1958. **127**(3301): p. 755–756. DOI: 10.1126/science.127.3301.755 88, 89

Özkaynak E., D.C. Rueger, E.A. Drier, C. Corbett, R.J. Ridge, and T.K. Sampath, et al., *OP-1 cDNA encodes an osteogenic protein in the TGF-ß family.* EMBO J Org, 1990. **9**(7): p. 2085–2093. 91, 92

Reddi A.H., (1984) *Extracellular matrix and development,* Extracellular Matrix Biochemistry, KA Piez and AH Reddi Editors. 1984, Elsevier: New York. p. 375. 84, 86, 89

Reddi A.H., (1997) *Bone morphogenesis and modeling: soluble signals sculpt osteosomes in the solid state.* Cell, 1997. **89**(2): p. 159–161. DOI: 10.1016/S0092-8674(00)80193-2 90, 94

Reddi A.H., (2000) *Morphogenetic messages are in the extracellular matrix: biotechnology from bench to bedside.* Biochem Soc Trans, 2000a. **28**(4): p. 345–349. DOI: 10.1042/0300-5127:0280345 86, 89, 92

Reddi A.H., (2000) *Morphogenesis and tissue engineering of bone and cartilage: inductive signals, stem cells, and biomimetic biomaterials.* Tissue Eng, 2000b. **6**(4): p. 351–359. DOI: 10.1089/107632700418074 86, 89, 91, 92, 94, 97, 101, 102

Reddi A.H., (2005) *BMPs: from bone morphogenetic proteins to body morphogenetic proteins.* Cytokine Growth Factor Rev, 2005. **16**(3): p. 249–250. DOI: 10.1016/j.cytogfr.2005.04.003 84, 97

Reddi A.H. and C.B. Huggins, (1972) *Biochemical sequences in the transformation of normal fibroblasts in adolescent rats.* Proc Natl Acad Sci USA, 1972. **69**(6): p. 1601–1605. DOI: 10.1073/pnas.69.6.1601 83, 86, 88, 89

Ripamonti U., (1988) *Paleopathology in Australopithecus africanus: a suggested case of a 3-million-year-old pre-pubertal periodontitis.* Am J Phys Anthropol, 1988. **76**(2): p. 197–210. DOI: 10.1002/ajpa.1330760208 84, 88

Ripamonti U., (1989) *The hard evidence of alveolar bone loss in early hominids of Southern Africa.* J Periodontol, 1989. **60**(2): p. 118–120. 84

Ripamonti U., Induction of cementogenesis and periodontal ligamentregeneration by bone morphogenetic proteins. In: Lindholm TS, editor. Bone Morphogenetic Proteins: Biology, Biochemistry and Reconstructive Surgery, Austin Academic Press;(1996). p.189–198. 99

Ripamonti U., Osteogenic proteins of the TGF-β superfamily. In: Henry H.L. and Norman A.W. editors. Encyclopedia of Hormones: (2003). p. 80–86. 86, 89, 91, 92, 94, 97, 101, 102

Ripamonti U., (2004) *Soluble, insoluble and geometric signals sculpt the architecture of mineralized bone.* J Cell Mol Med, 2004. **8**: p. 169–180. DOI: 10.1111/j.1582-4934.2004.tb00272.x 103

Ripamonti U., (2006) *Soluble osteogenic molecular signals and the induction of bone formation.* Leading Opinion Paper, Biomaterials, 2006. **27**(6): p. 807–22. DOI: 10.1016/j.biomaterials.2005.09.021 90, 91, 94, 96

Ripamonti U., (2007) *Recapitulating development: a template for periodontal tissue engineering.* Tissue Eng, 2007. **13**(1): p. 51–71. DOI: 10.1089/ten.2006.0167 86, 91, 92, 94, 96, 101, 102

Ripamonti U. and J-C. Petit, (1991) *Évidence de maladies pariodontales chez les Australopithèques rèvèlèe par l'examen de restes osseux fossilisès.* L'Anthropologie, 1991. **95**: p. 391–400. 84

Ripamonti U., and J-C. Petit, (2009) *Bone morphogenetic proteins, cementogenesis, myoblastic stem cells and the induction of periodontal tissue regeneration.* Cytokine Growth Factor Rev. 2009, **20**(5–6): p. 489-499. DOI: 10.1016/j.cytogfr.2009.10.016 101, 102

Ripamonti U., and A.H. Reddi, (1994) *Periodontal regeneration: Potential role of bone morphogenetic proteins.* J Perriodont Res, 1994, **29**: p. 225–235. DOI: 10.1111/j.1600-0765.1994.tb01216.x 96, 97, 101

Ripamonti U., and A.H. Reddi, (1997) *Tissue engineering, morphogenesis, and regeneration of the periodontal tissues by bone morphogenetic proteins.* Crit Rev Oral Biol Med, 1997. **8**(2): p. 154–163. DOI: 10.1177/10454411970080020401 91, 96

Ripamonti U., S. Ma, N.S. Cunningham, L. Yeates and A.H. Reddi, (1992) *Initiation of bone regeneration in adult baboons by osteogenin, a bone morphogenetic protein.* Matrix, 1992. **12**(5): p. 369–380. 90, 91

Ripamonti U., S.S. Ma, N.S. Cunningham, L. Yeates, and A.H. Reddi, (1993) *Reconstruction of the bone-bone marrow organ by osteogenin, a bone morphogenetic protein and demineralized bone matrix in calvarial defects of adult primates.* Plast Reconstr Surg, 1993. **91**(1): p.27–36. DOI: 10.1097/00006534-199301000-00005 90, 91

Ripamonti U., M. Heliotis, B. van den Heever, and A.H. Reddi, (1994) *Bone morphogenetic proteins induce periodontal regeneration in the baboon (*Papio ursinus). J Periodontal Res, 1994. **29**(6): p. 439–445. DOI: 10.1111/j.1600-0765.1994.tb01246.x 91, 92, 94, 102

Ripamonti U., M. Heliotis, D.C. Rueger, and T.K. Sampath, (1996) *Induction of cementogenesis by recombinant human osteogenic protein-1 (hOP-1/BMP-7) in the baboon.* Archs Oral Biol, 1996. **41**(1): p. 121–126. DOI: 10.1016/0003-9969(95)00110-7 97, 99

Ripamonti U., N. Duneas, B. van den Heever, C. Bosh, and J. Crooks, (1997). *Recombinant transforming growth factor-β₁ induces endochondral bone in the baboon and synergizes with recombinant osteogenic protein-1 (bone morphogenetic protein-7) to initiate rapid bone formation.* J Bone Miner Res, 1997. **12**(10): p. 1584–1595. DOI: 10.1359/jbmr.1997.12.10.1584 102

Ripamonti U., B. van den Heever, J. Crooks, M.M. Tucker, T.K. Sampath, D.C. Rueger and A.H. Reddi, (2000) *Long-term evaluation of bone formation by osteogenic protein 1 in the baboon and relative efficacy of bone-derived bone morphogenetic proteins delivered by irradiated xenogenic collagenous matrices.* J Bone Miner Res, 2000a. **15**(9): p. 1798–1809. DOI: 10.1359/jbmr.2000.15.9.1798 86, 91, 97

Ripamonti U., J. Crooks, T. Matsaba, and J. Tasker, (2000) *Induction of endochondral bone formation by recombinant human transforming growth factor-β2 in the baboon (*Papio ursinus). Growth Factors, 2000b. **17**(4): p. 269–285. DOI: 10.3109/08977190009028971 102

Ripamonti U., J. Crooks, J-C. Petit, and D. Rueger, (2001) *Periodontal tissue regeneration by combined applications of recombinant human osteogentic protein-1 and bone morphogenetic protein-2. A pilot study in Chacma baboons (*Papio ursinus). Eur J Oral Sci, 2001. **109**(4): p. 241–248. DOI: 10.1034/j.1600-0722.2001.00041.x 97, 99, 101

Ripamonti U., J. Crooks, J. Teare, J-C. Petit, and D.C. Rueger (2002), *Periodontal tissue regeneration by recombinant human osteogenic protein-1 in periodontally-induced furcation defects of the primate Papio ursinus.* S Afr J Sci, 2002. **98**(2): p. 361–368. DOI: 10.1111/j.1600-0765.2007.00987.x 99

Ripamonti U., L.N. Ramoshebi, J. Patton, T. Matsaba J. Teare, and L. Renton, (2004) *Soluble signals and insoluble substrata: novel molecular cues instructing the induction of bone.* In: Massaro EJ, Rogers JM, editors. *The Skeleton.* Totowa: Humana Press; 2004. p. 217–227. 97, 99, 101, 102

Ripamonti U., N.N. Herbst, and L.N. Ramoshebi, (2005) *Bone morphogenetic proteins in craniofacial and periodontal tissue engineering: experimental studies in the non-human primate Papio ursinus.* Cytokine Growth Factor Rev, 2005. **16**(3): p. 357–368. DOI: 10.1016/j.cytogfr.2005.02.006 91, 92

Ripamonti U., C. Ferretti, and M. Heliotis, (2006) *Soluble and insoluble signals and the induction of bone formation: molecular therapeutics recapitulating development.* J Anat, 2006a. **209**(4): p. 447–468. DOI: 10.1111/j.1469-7580.2006.00635.x 86, 88, 89, 91, 94, 97

Ripamonti U., J. Teare, and J-C. Petit, (2006) *Pleiotropism of bone morphogenetic proteins: from bone induction to cementogenesis and periodontal ligament regeneration.* J Int Acad Periodontol, 2006b. **8**(1): p. 23–32. 97, 99

Ripamonti U., M. Heliotis, and C. Ferretti, (2007) *Bone morphogenetic proteins and the induction of bone formation: from laboratory to patients.* Oral Maxillofac Surg Clin North Am, 2007. **19**(4): p. 575–589. DOI: 10.1016/j.coms.2007.07.006 101

Ripamonti U., L.N. Ramoshebi, J. Teare, L. Renton, and C. Ferretti, (2008) *The induction of endochondral bone formation by transforming growth factor-β3: Experimental studies in the non-human primate* Papio ursinus.J Cell Mol Med, 2008. **12**(3): p. 1029–1048. DOI: 10.1111/j.1582-4934.2008.00126.x 102

Ripamonti U., J-C. Petit, and J. Teare, (2009) *Cementogenesis and the induction of periodontal tissue regeneration by the osteogenic proteins of the transforming growth factor-ß superfamily.* J Priodont Res, 2009a. **44**(2): p. 141–152. DOI: 10.1111/j.1600-0765.2008.01158.x 91, 92, 94, 96, 101, 102

Ripamonti U., C. Ferretti, Teare J., and L. Blann, (2009) *The transforming growth factor-β isoforms and the induction of bone formation: Implications for reconstructive craniofacial surgery.* J Craniofac Surg, 2009b. **20**(5): p. 1544–1555. DOI: 10.1097/SCS.0b013e3181b09ca6 102

Ripamonti U., R. Parak, and J-C. Petit, (2009) *Induction of cementogenesis and periodontal ligament regeneration by recombinant human transforming growth factor-ß3 in matrigel with rectus abdominis responding cells.* J Periodont Res, 2009c. **44**(1): p. 81–87. DOI: 10.1111/j.1600-0765.2008.01086.x 102

Sacerdotti C, and G. Frattin, (1901) *Sulla produzione eteroplastica dell'osso.* R Accad Mead Torino, 1901. p. 825–836. 88, 89

Sampath T.K., and A.H. Reddi, (1981) *Dissociative extraction and reconstitution of extracellular matrix components involved in local bone differentiation.* Proc Natl Acad Sci USA, 1981. **78**(12): p. 7599–7603. DOI: 10.1073/pnas.78.12.7599 89, 90, 91

Sampath T.K., and A.H. Reddi, (1983) *Homology of bone-inductive proteins from human, monkey, bovine and rat extracellular matrix.* Proc Natl Acad Sci USA, 1983. **80**(21): p. 6591–6595. DOI: 10.1073/pnas.80.21.6591 89, 90

Sampath T.K., J.C. Maliakal, P.V. Hauschka, W.K. Jones, H. Sasak, and R.F. Tucker, et al., (1992) *Recombinant human osteogenic protein-1 (hOP-1) induces new bone formation in vivo with a specific activity comparable with natural bovine osteogenic protein and stimulates osteoblasts proliferation and differentiation in vitro.* J Biol Chem, 1992. **267**(28): p. 20352–20362. 91

Seo B-M., M. Miura, S. Gronthos, P.M. Bartold, S. Batouli, and J. Brahin, et al., (2004) *Investigation of multipotent postnatal stem cells from human periodontal ligament.* Lancet, 2004. **364**(9429): p. 149–155. DOI: 10.1016/S0140-6736(04)16627-0 101

Sigurdsson T.J., M.B. Lee, K. Kubota, T.J. Turek, J.M. Wozney, and U.M.E. Wikesjö, (1995) *Periodontal repair in dogs: recombinant human bone morphogenetic protein-2 significantly enhances periodontal regeneration.* J Periodontol, 1995a. **66**(2): p. 131–138. 101

Sigurdsson T.J., D.N. Tatakis, M.B. Lee, and U.M.E. Wikesjö, *Periodontology regenerative potential of space-providing expanded polytetrafluoroethylene membranes and recombinant human bone morphogenetic proteins.* J Periodontol, 1995b. **66**(6): p. 511–521. 101

Teare J.A., L.N. Ramoshebi, and U. Ripamonti, (2008) *Periodontal tissue regeneration by recombinant human transforming growth factor-β3 in Papio ursinus.* J Periodont Res, 2008. **43**(1): p. 1–8. DOI: 10.1111/j.1600-0765.2007.00987.x 102

Thesleff I. and P. Sharpe, (1997) *Signalling networks regulating dental development.* Mech Dev, 1997. **67**(2): p. 111–123. DOI: 10.1016/S0925-4773(97)00115-9 91, 97

Thomadakis G., L.N. Ramoshebi, J. Crooks, C.D. Rueger, and U. Ripamonti, (1999) *Immunolo-calization of bone morphogenetic protein-2 and -3 and osteogenic protein-1 during murine tooth root morphogenesis and in other craniofacial structures.* Eur J Oral Sci, 1999. **107**(5): p. 368–377. DOI: 10.1046/j.0909-8836.1999.eos107508.x 91, 92, 97

Trueta J. *The role of the vessels in osteogenesis.* J Bone Joint Surg, 1963. **45B**: p. 402–418. 94, 96

Turing A.M., (1952) *The chemical basis of morphogenesis.* Philos Trans Roy Soc Lond, 1952. **237**(7): p. 37–41. DOI: 10.1098/rstb.1952.0012 86

Urist M.R., (1965) *Bone: Formation by autoinduction.* Science, 1965. **150**(698): p. 893–899. DOI: 10.1126/science.150.3698.893 83, 86, 88, 89

Urist M.R., B.F. Silverman, K. Buring, F.L. Dubuc, and J.M. Rosenberg, (1967) *The bone induction principle.* Clin Orthop, 1967. **53**(83): p. 243. 86, 88, 89, 90, 102

Urist M.R., T.A. Dowell, P.H. Hay, and B.S. Strates, (1968) *Inductive substrates for bone formation.* Clin Orthop and Relat Res, 1968. **59**: p. 59–96. 86, 88, 89

Urist M.R., and B.S. Strates, (1971) *Bone morphogenetic protein.* J Dent Res, 1971. **50**(6): p. 1392–1406. 86, 89

Vainio S., I. Karanova, A. Jowett, and I. Thesleff, (1993) *Identification of BMP-4 as a signal mediating secondary induction between epithelial and mesenchymal tissues during early tooth development.* Cell, 1993. **75**(1): p. 45–58. DOI: 10.1016/S0092-8674(05)80083-2 91, 97

Willestaedt H., G. Levander, and L. Hult, (1950) *Studies in osteogenesis.* Acta Orthop Scand, 1950. **19**(4): p. 419–432. DOI: 10.3109/17453675008991101 88, 89

Wozney J.M., V. Rosen, A.J. Celeste, L.M. Mitsock, M.J. Whitters, R.W. and Kriz, et al., (1988) *Novel regulators of bone formation; molecular clones and activities.* Science, 1988. **242**(4885): p. 1528–34. DOI: 10.1126/science.3201241 91

Zeichner-David M., (2000) *Regeneration of periodontal tissues: cementogenesis revisited.* Periodontol 2000, 2006. **41**:196–217. DOI: 10.1111/j.1600-0757.2006.00162.x

Zhao Z.M., J.E. Berry, and M.J. Somerman, (2003) *Bone morphogenetic protein-2 inhibits differenti-ation and mineralization of cementoblasts in vitro.* J Dent Res, 2003. **82**(1): 101

CHAPTER 4

Dynamics for Pulp-Dentin Tissue Engineering in Operative Dentistry

Dimitrios Tziafas, DDS, PhD

4.1 CHAPTER SUMMARY

The ability of tissue engineering approaches to stimulate regeneration of the diseased dental tissues offers exciting opportunities for the future. This chapter reviews the literature-based concepts on biological mechanisms underlying the development, function, regeneration potential of the pulp-dentin complex, and how they can open new directions to the scientists to devise more realistic therapeutic strategies for the treatment of the dental diseases. It is important to realize that the nature and specificity of the biological mechanisms by which the traumatized pulp-dentin complex is therapeutically healed determine the properties of the newly formed matrix and play a critical role in the outcome of dental treatment. The dental pulp cells are genetically programmed to become potential preodontoblasts able to differentiate into dentin- or dentin-like matrix-forming cells. Expression of the dentinogenic potential requires specific epigenetic signals, which have been investigated in numerous ex vivo and in vivo models. Knowledge of the most important signaling mechanisms, provided by the dentin-pulp complex microenvironment in health and disease, has set the stage for dental tissue engineering and regeneration.

4.2 INTRODUCTION

The nature and specificity of the biological mechanisms by which the traumatized pulp-dentin complex is healed determine the properties of the new tissue and play a critical role in the outcome of dental treatment. The dental pulp cells are genetically programmed to become dentin- or dentin-like matrix-forming cells. Differentiation requires specific epigenetic signals, which have been investigated in numerous ex vivo and in vivo models. The most important signaling mechanisms, provided by the dental tissues' microenvironment in pulp-dentin complex development, structure, and function in health and disease, in controlling therapeutic regeneration the pulp-dentin complex regeneration are discussed.

4.3   DYNAMICS FOR PULP/DENTIN TISSUE ENGINEERING IN OPERATIVE DENTISTRY

In 1997, Michael Barnett wrote that *'when I consider the mode of dental practice we learned in dental schools 30 years ago and compare it with today's practice, I see great process but not a quantum leap. Work currently underway suggests that a quantum leap is possible and is, in fact, achieavable. Those individuals lucky enough to be practicing well into the 21st century will have the opportunity to utilize the fruits of current research.'* The challenge of the today dental research is to integrate advanced biological knowledge into the clinical approach to the problems of dental practice. Especially, advances in molecular biology and bioengineering research is now integrated into the clinical problems of dentistry. Our basic concepts on biological mechanisms underlying the development, function, and regeneration potential of the dental tissues, more particularly of the skeletal body of the tooth named pulp-dentin complex, have already opened new directions to the scientists to devise more realistic therapeutic strategies for the treatment of the dental diseases. These concepts are reviewed in the present chapter.

4.3.1   PULP/DENTIN COMPLEX IN HEALTH

4.3.1.1   Structural and Embryological Aspects

The pulp and dentin form an embryological and functional entity and are widely considered as a complex. Dentin is a mineralized, extracellular matrix that forms the skeletal tissue of the tooth. The dental pulp is entirely enclosed by dentin in pulp chamber and root canal(s) of the tooth.

The dental pulp is a well vascularized, specialised connective tissue derived from oral ectomesenchyme (Fig. 4.1 A). It contains cells which belong to different groups that provide odontogenic, defensive, nutritive, and sensory functions (odontoblasts, fibroblasts, undifferentiated progenitor cells, and defense cells), extracellular matrix (collagen and reticular fibres), blood vessels, and nerves. Pulp structure is not uniform, consisting of the odontogenic pulp periphery and the pulp proper (Fig. 4.1 B).

Three border zones can be distinguished in the coronal and the substantial portion of the radicular tissue, forming the odontogenic pulp region:

(i) Odontoblast layer. It is a pseudostratified layer of highly differentiated cells responsible for the formation of the dentin during embryonic tooth development and its repairing during the life span of the pulp organ. Odontoblasts have columnar cell bodies and odontoblastic procesess that extend into the nearby dentinal tubules.

(ii) Cell-free zone. It is an approximately 40μm subodontoblastic zone, more distinct in the coronal pulp, which contains numerous branching cytoplasmic processes from cells located in the adjacent cell-rich zone, the major portion of the sub-odontoblastic capillary plexus and the terminal branches of the sensory and autonomic nerve fibres.

(iii) Cell-rich zone. It contains bipolar cells (fibroblasts and undifferentiated cells) with spindle-shaped nuclei arranged with their cytoplasmic processes perpendicularly or parallelry to the dentin in the coronal and radicular pulp, respectively. On the inner side, cell-rich zone is continuous with the central pulp parenchyma.

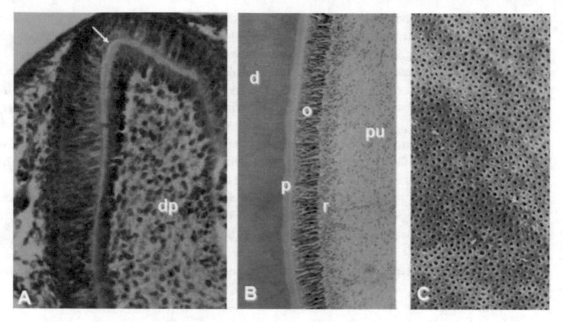

Figure 4.1: A. Mouse molar tooth germ. Initiation of primary dentinogenesis (arrow) and differentiation of odontoblasts from the cells of ectomesenchymal dental papilla. B. Human premolar immature tooth. Primary dentin (d), predentin (p), odontoblast layer (o), cell-rich zone (cr and pulp proper (pu). C. Human premolar mature tooth. Tubular structure of primary dentin.

The central pulp proper consists mainly by fibroblasts and undifferentiated cells, extracellular matrix, large blood vessels, and nerve trunks. Pulp tissue is rich in collagenous fibrils, fibers and thin fiber bundles. Approximately 30-45% of the pulpal collagen fibres consist of type III collagen, while type I collagen is also present in the pulp. All the pulp elements are embedded in a gel-like ground substance with a high water content containing chondroitin sulfate, hyaluronic acid, dermatan sulfate, proteoglycans, and glycoproteins. Vessels enter and exit the pulp through the apical foramen. Narrow arteries and thin-walled arterioles run through the center of the radicular pulp and show an extensive branching in the periphery of the coronal pulp, forming the sub-odontoblastic capillary plexus. Venules and larger veins run alongside the arteries in a spiralling course. Arteriovenous anastomoses exist in the coronal and the radicular pulp, independent of the peripheral pulp capillaries. Myelinated and unmyelinated nerve fibres enter the pulp through the apical foramen, running in close proximity with the blood vessels. Myelinated fibres form several branchings in the pulp periphery where they, finally, lose their myelin sheaths.

The dental pulp originates from the oral ectomesenchyme. During tooth development, a cluster of ectomesenchymal cells, including oral mesenchymal cells of the first branchial arch and neural crest cells, named dental papilla, is formed within developing dental epithelium (enamel organ). The boundary between enamel organ and dental papilla is the basement membrane. When the enamel organ reaches the late bell stage of its development, deposition of hard tissue begins. Cells in the epithelial side of the basement membrane become ameloblasts forming tooth enamel, while the rest of the organ progressively disappears. Cells in the mesenchymal side of the basement membrane become odontoblasts forming primary dentin. The layer of odontoblasts and the remaining mass of dental papilla, consisting of spindle-shaped undifferentiated cells and a ground substance, is transformed into dental pulp, surrounded now by dentin. This transformation is characterised by the differentiation of most of the undifferentiated cells into active fibroblasts and the gradually increasing volume of collagen fibers. Some of the original undifferentiated cells remain in the dental pulp as reservoir, assuming at a later time odontogenic or defensive functions.

Only the dental pulp cells possess the ability to differentiate into odontoblasts. It is generally accepted that no other population of mesenchymal cells is able to differentiate into odontoblast or odontoblast-like cells, and this specific ability seems to be acquired by morphogenic influences given during tooth development. Interactions between the oral ectoderm and the adjacent ectomesenchymal cells, during the initial events of tooth formation, seem to determine the odontoblastic potential of dental papilla cells (Thesleff and Vaaitokari, 1992; Ruch et al., 1995).

4.3.1.2 Odontoblasts and Dentin Matrix

The nature and specificity of the dontoblastic cell lineage and the associated biosynthetic activity (dentinogenesis) are the most critical issues in the dentin/pulp complex, in terms of regenerative capacity of the tooth. Odontoblasts are elongated polarized cells, aligned in a single layer at the periphery of the dental pulp; they are highly differentiated cells expressing unique functional characteristics, incapable of further cell division (Stanley and Lundy, 1972; Chiba et al., 1967; Holz and Baume, 1973). Odontoblasts originate from the neural crest cells, and their differentiation results from continuous reciprocal interactions between epithelial and mesenchymal components of the developing tooth germ (Slavkin et al, 1981; Thesleff and Hurmerinta, 1981; Kollar EJ, 1983; Ruch JV, 1985). Expression of odontoblast phenotype initiating by withdrawal from the cell cycle is characterized by a sequence of cytological and functional changes which occur to each tooth casp according to a specific temporo-spatial pattern (Moullec N, 1978; Ruch JV, 1985).

Morphologically, odontoblasts are characterised as elongated cells having a basal nuclear position and one cytoplasmic process that is extended into the tubular primary dentin (Tominaga et al., 1984). The most important criterion for characterization of functional odontoblast differentiation is the secretion of a tubular matrix by elongated polarized cells. However, odontoblast differentiation implies also transcriptional and traslational modifications, enabling the cells to deposit dentin components. Fully differentiated odontoblasts synthesize and mainly secrete collagen proteins; therefore, a unique set of non-collagenous proteins, such as the dentin-specific proteins, and proteins frequently found in a variety of tissues are synthesized (see in Veis A, 1985; Linde A, 1989; Bronckers et al., 1989; Robey B, 1989; D'Souza et al., 1992; Linde and Goldberg, 1993; Veis A, 1993; George et al., 1994; Butler and Ritchie, 1995; Goldberg and Lasfargues, 1995).

The collagen proteins consist 90% of the total organic dentin matrix, and they are composed mainly by type I (95%), type V (3%) and type VI (minor components), while the presence of type I trimmer and type III is still controversial (Butler W, 1972; Lesot et al., 1981; Butler W, 1984). The group of glycoproteins includes dentin sialoprotein, osteonectin, osteopontin and bone sialoprotein (Butler et al., 1992; Bronckers et al., 1989; Ritchie et al., 1995). Dentin sialoprotein is a sialic, acid-rich, acidic glycoprotein expressed primarily or exclusively by odontoblasts and related odontoblast-like cells (Butler et al., 1981, 1992; Ritchie et al., 1995; D'Souza et al., 1995). The phosphoproteins consist over the 50% of the total amount of the non-collagenous proteins, including two highly phosphorylated molecules, one moderately and one weakly phosphorylated molecule and a serine-rich phosphoprotein known as dentin matrix protein (Veis and Perry, 1967; Butler et al., 1983; George et al., 1993). Other molecular components of dentin are osteocalcin (Bronckers et al., 1989; Gorter de Vries et al., 1986), lipids that constitute 0.35% of the whole tissue and (Goldberg et al., 1995), serum derived proteins (Kinoshita Y, 1979; Butler et al., 1981; Takagi et al., 1990) and growth factors, which have attracted recent attention. This polypeptide group includes transforming growth factor-beta molecules (TGF-betas) and insulin-like growth factor molecules/IGFs (Finkelman et al., 1990), bone mophogenetic proteins/BMPs (Bessho et al., 1991), and a rat incisor dentin polypeptide different from TGF-beta and BMP, expressing an in vitro chondrogenic activity (Amar et al., 1991). TGF-beta 1, 2, 3 have been found as different extracellular matrix compartments in dentin (Smith et al., 2001). An amount of these isoforms appear to be tightly associated with the collagen; they are not released from the dentin matrix by the demineralization process requiring collagenase digestion.

Three types of dentin can be distinguished:

Primary dentin. The hard tissue that constitutes the skeletal body of each tooth, produced absolutely by primary odontoblasts at a high rate, during the embryonic period of tooth development as mantle dentin, consisting of the outermost 10 to 30μm dentin layer, and circumpulpal dentin, representing the bulk of the dentin mass up to the end of root formation (Fig. 4.1 A,B, and C).

Secondary dentin. Dentin, which is formed at a very reduced rate, post-developmentally, by primary odontoblasts or their replacement cells, throughout the life span of the dental pulp organ.

Tertiary dentin. The dentin-like matrix which is formed as a response to exogenous irritation of the dentin-pulp complex or as a result of the effects of biologically active molecules on the dental pulp cells. It must be clear that therapeutic regulation of dentin formation in traditional or regenerative approaches has as a result formation of tertiary dentin. It has been generally accepted that the generation of odontoblastic cell lineage shares common chraracteristics in both primary and tertiary dentinogenesis.

In order to approach therapeutic regulation of dentin formation, the understanding of molecular and cellular mechanisms involved in dentinogenesis, during tooth formation and dental pulp repair, are of great importance. Hundreds of investigations described the complex genesis of dentin, either the generation of the specific cells or the interrelated processes of organic matrix synthesis, exocytosis, endocytosis and mineral formation (Baume LJ, 1980; Linde and Goldberg, 1993). Understanding how generation of odontoblastic cell lineage is achieved during tooth development is a fundamental issue of current interest to scientists working in regenerative dentistry.

4.3.1.3   Generation of Odontoblastic Cell Lineage During Tooth Development

During tooth morphogenesis, ectomesenchymal cells of the dental papilla, progressively acquire increasing levels of specification, and as a consequence, they become the only cells which respond to signals from the enamel epithelium by differentiating into odontoblasts (Fig. 4.1 A). Thesleff et al., 1995 described the most important molecular features that regulate tooth specificity:

 (i) Morphogens from the family of bone morphogenetic proteins (BMPs), BMP-2 and BMP-4 regulate expression of the homeobox containing genes Msx-1 and Msx-2 (MacKenzie et al., 1991, 1992; Satokata and Mass, 1994) and transcription factors, such Lef 1 (Oosterwegel et al., 1993; Van Genderen et al., 1994).

 (ii) Many genes and transcription factors express molecules at the cell surface and the extracellular matrix, such as syndecan-1 (Thesleff et al., 1987; Vainio et al., 1989; Bernfield et al., 1992) and tenascin (Vainio et al., 1989; Erickson H, 1993).

 (iii) Condensation of tooth mesenchyme contributes to epithelial morphogenesis by regulation of proteolytic enzymes, growth factors such as FGFs and EGF and their receptors (Partanen and Thesleff, 1987; Kronmiller et al., 1991; Niswander and Martin, 1992; Jernvall et al., 1994; Kettunen and Thesleff, 1998).

 (iv) Epithelial morphogenesis lead to an extensive remodeling of epithelial-mesenchymal interface (basement membrane) in a tooth-specific pattern and progressively regulates the dental mesenchyme environment.

 (v) It has been repeatedly demonstrated that odontoblast terminal differentiation can only occur under epigenetic signals, given by stage-specific inner dental epithelium and basement membrane or by substituting network of matrix molecules and growth factors. Matrix molecules, growth factors, transcription factors and their specific receptors have shown to represent a network of molecular mechanisms leading peripheral papilla ectomesenchymal cells to express their competence, the odontoblast phenotype (Ruch JV, 1985; Lesot et al., 1981; Thesleff et al., 1981; Slavkin et al, 1981; Kubler et al., 1988). Preodontoblasts, which were mechanically dissociated from the inner epithelium after their last mitosis, became fully differentiated cells only in association with the basement membrane. Odontoblast polarization and initiation of predentin secretion might require specific structural and compositional organization of the matrix in the basement membrane (Ruch JV, 1985); it appears that trapped molecules rather than permanent components of the basal lamina play the crucial role (Ruch et al., 1995).

The process of odontoblast differentiation has been desrcibed as a cascade of 3 cytological . steps:

(i) Cessation of cell division. Odontoblasts are post-mitotic cells incapable of further cell division (Chiba et al., 1967; Holz and Baume, 1973). During the last mitosis of preodontoblasts, the mitotic spindle lies perpendicular to the basement membrane (Osman and Ruch, 1976). Daughter cells in close proximity to the basement membrane become odontoblasts apical cells, named Holl's cells, constitute a layer of potential preodontoblastic cell lineage, giving rise to replacement odontoblast-like cells (Baume LJ, 1980).

(ii) Cell polarization. The peripheral post-mitotic preodontoblasts become larger and elongate, their nuclei take up a position in the apical part of the cell body and a trunk of cytoplasmic processes is formed. The cisternae of the granular endoplasmic reticulum progressively become parallel to the long axis of the cells. Further enlargement of Golgi region, specific modifications of the plasma membrane and cytoskeletal rearrangement characterize odontoblast polarization at the ultrastructural level (Tominaga et al., 1984).

(iii) Predentin formation. Polarized cells synthesize and secrete extracellular matrix in a polar pattern. Secretion of predentin around the odontoblastic process gives a tubular appearance in the forming matrix.

The most characteristic changes in the matrix composition during odontoblast differentiation are the disappearance of collagen type III, the increasing of the proteoglycan tenascin, the progressive accumulation of the glycoprotein fibronectin, and the proteoglycan decorin around the most apical pole of polarizing cells (Lesot et al., 1981; Thesleff and Hurmerinta, 1981; Thesleff et al., 1987; Meyer et al., 1989; Yoshiba et al., 1994). A functional network of matrix molecules in the basement membrane, such as the heparin/fibronectin and the following epithelially-derived growth factors seems to control the biological process of odontoblast diferentiation, while further synergistic interactions with other endogenous and circulating growth factors could not be excluded.

In more detail, the expression of some bFGFs during tooth development was confined to dental epithelial cells, while they are also coexpressed in dental mesenchyme at stages when epithelial - mesenchymal signalling regulates tooth morphogenesis; they act in networks with other signal molecules (Thesleff et al., 1995). TGF-betas have been implicated as important regulatory factors for odontoblast differentiation during tooth development; they are synthesized by preameloblasts and progressively trapped and activated in the basement membrane, influencing complete expression of odontoblast phenotype (Cam et al., 1990; D'Souza et al., 1990; Wise and Fan, 1991; Vaahtokari et al., 1991; Jepsen et al., 1992; Thesleff and Vaahtokari, 1992; Heikinheimo et al., 1993). The expression of TGF-betas by odontoblasts leads to their sequestration within the dentine matrix, where they are available to participate in tissue homeostasis and reparative dentinogenesis after pulp amputation (Finkelman et al., 1990; Cassidy et al., 1997). Expression of BMPs during primary dentinogenesis has been identified; BMP-mRNA was found in dental papilla and functional odontoblasts (Lyons et al., 1990; Heikinheimo K, 1994), BMP-4 in early tooth morphogenesis (Vainio et al., 1993; Heikinheimo K, 1994) and BMP-6 in induction of odontoblast cell lineage (Heikinhcimo K, 1994). Extracts with BMP-like activity were isolated from dentin (Butler et al., 1992; Katz and Reddi, 1988; Kawai et al., 1989; Bessho et al., 1991). Investigations concerning the expression of neurotrophins and their receptors in developing teeth, suggest apart from their neuronal functions, a regulatory role during proliferation of the cells in the inner dental epithelium and at the onset of odontoblast differentiation (Byers et al., 1990; Mitsiadis et al., 1992, 1993; Mitsiadis and Luukko, 1995). Since NGF induces transcription of TGF-beta1 (Kim S, 1990), it is not unreasonable to postulate that these molecules, identified also during pulp healing (Byers et al., 1992), can activate expression of TGF-beta1 in differentiating odontoblasts during both development and repair. The distribution pattern of IGF-I and its receptor, as has been investigated by immunohistochemistry, showed first expression in developing teeth in already polarized odontoblasts, strong expression in predentin and subsequent expression in pre-ameloblasts undergoing differentiation (Joseph et al., 1993, 1994; Young WG, 1995).

4.3.2   PULP/DENTIN COMPLEX IN DISEASE

4.3.2.1   General Pathological Aspects

The pathology of dental pulp represents a network of inflammatory reactions of pulpal cells, microcirculation and nerves whenever dentin and pulp is affected by caries, restorative procedures and trauma (Fig. 4.2 A).

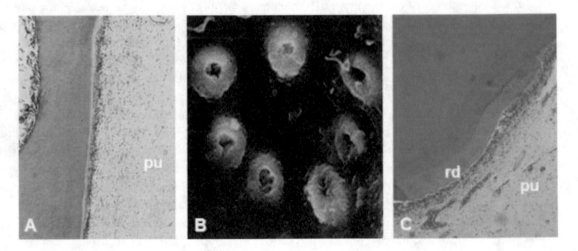

Figure 4.2: A. Mechanically-prepared dentinal cavity in a dog tooth. Normal architecture of the underlying periphery and pulp proper (pu) after application of a calcium-hydroxide-based material. B. Formation of peritubular dentin in a carious human tooth. C. Formation of reactionary tertiary dentin (rd) after transdentinal bio-active influence.

Specific structural and functional aspects directly affect the outcome of the fundamental defensive mechanisms in the dental pulp:

(i) The peripheral odontoblast palisade act as a selective barrier regulating the transfer of ions, molecules or fluids from pulp periphery to the proper pulp (Koling and Rask-Andersen, 1984; Bishop MA, 1985). The tight junctions in the apical zone of the odontoblast cell bodies might be the main mechanism for this control function, allowing only a small portion of tissue products or serum molecules to affect the pulp (Turner et al., 1989).

(ii) The most significant difference between the pulp and other connective tissues in the patho-physiology of tissue disorder is the low compliance environment of the pulp organ. The initial vascular reactions during pulp inflammation (vasodilatation and increased vessel permeability) taking place in the rigid enclosed pulp chamber, create conditions of increased hydrostatic tissue pressure. Despite the fact that the pressure increase is a local phenomenon (Van Hassel HJ, 1971) and negative feedback mechanism control changes in hydrostatic pressure by the oedema-preventing mechanisms (Heyeraas KJ, 1990); dental pulp pressure can quickly suffer irreversible damage. In the low compliance environment of the pulp chamber, tissue responds to vasoconstriction in a similar way to that of other systems. On the contrary, pulp response to vasodilating agents, such as the mediators released during initial tissue insult by deep caries or mechanical trauma, is very different. In the pulp system vasodilatation resulting in a sharp, transient increase in pulpal blood flow can be followed by a dramatic vasoconstriction (Kim S, 1990). On the other hand, osmotic feedback mechanism, in which increased filtration normally tends to dilute fluid proteins reducing colloid osmotic pressure is not effective in the low-compliance systems because dilution is not possible owing to the relatively constant tissue volume (Heyeraas KJ, 1990). Dental pulp healing is not always follow the sequence of events taking place normally in other connective tissues. Since pulp repair is strongly dependent on a number of factors, exacerbation of an initial trauma very often leads to general tissue necrosis.

(iii) The activation of sensory nerve fibres during various dental procedures does not only serve to signal pain but also releases vasoactive peptides (substance P, CGRP, NKA, VIP, and NPY), leading to vascular reactions locally within the pulp environment (Akai and Wakisaka, 1990; Olgart LM, 1990). Such local reflex reactions might be beneficial to the pulp organ under low-grade tissue irritations (Olgart LM, 1990). The stimulation of periodontal nerve fibres can also increase blood flow in the pulp (Olgart LM, 1992). The stimulation of sensory nerve fibres may participate in control of the function of arteriovenous anastomoses, regulating blood supply and, subsequently, tissue healing in the pulp (Heyeraas and Kvinnsland, 1993).

(iv) The existence of specific pulp environment, i.e., the totally enclosed pulp tissue by dentinal walls, which are aligned by the border odontoblast palisade, is the most important requirement for the survival of dental pulp tissue. Whenever the basic structure of dentin-odontoblast layer is affected due to exogenous irritations, cells of underlying pulp mesenchyme possess the ability to replace denegerated odontoblasts or to differentiate into new hard tissue-forming cells (see in Baume LJ, 1980; Yamamura T, 1985; Lesot et al., 1993, 1994; Tziafas D, 1994). It might be suggested that the expression of the above mentioned specific potential of the pulp cells depends on dynamic interactions between type of trauma, reactions of the neurovascular system within the pulp microenvironment and structural\functional state of the pulp tissue. Non-inflamed dental pulp could be characterized as an appropriate environment, which is favorable for tissue healing allowing expression of the intrinsic potential of pulp cells.

(v) During the caries process, the dentinal matrix is demineralized and progressively invaded by micro-organisms. Bacterial metabolites and products released from hydrolysed and enzymatically digested dentin affect the underlying pulp (Bergenholtz G, 1981; Trowbridge HO, 1981; Larmas M, 1986). Under initial lesions, bacteria do not involve the dentinal tubules, while only slight cytological modifications of the odontoblasts (reduction in size and number of intracellular organelles, swollen mitochondria, enlargement of the intercellular spaces between odontoblasts, etc.) can be detected (Brannstrom and Lind, 1965; Baume LJ, 1980; Magloire et al., 1981; Yoshida and Massler, 1984; Couve E, 1986; Silverstone and Mjor, 1988). The increase of protein synthesis by odontoblasts in the affected area, leading to intratubular mineralization (dentin sclerosis), is the main pulp response to the superficial carious lesions. However, under chronic carious processes, irritating factors from bacteria and tissue breakdown affect odontoblasts for long periods and progressively destroy them. Capillaries leak plasma, and inflammatory cells (neutrophils, histiocytes and monocytes) become extravasated, releasing various chemotactic factors and cytokines. Prolonged release of bacterial metabolites and products to the proper pulp develop chronic inflammatory reactions and total pulp necrosis (Bergenholtz G, 1977; Trowbridge HO, 1981; Langeland K, 1987).

(vi) The mechanico-chemical irritation due to cavity preparation (Brannstrom M, 1961; Zach L, 1972), dentin dehydration (Langeland K, 1959; Brannstrom M, 1963), acute or chronic physical traumas (Andreasen JO, 1970; Jacobsen and Zachrisson, 1975; Meister et al., 1980; Andreasen and Andreasen, 1994) and chemical injuries (Langeland et al., 1971; Mjor IA, 1977; Brannstrom and Olivera, 1979; Cox et al., 1982; Stanley HR, 1993) can cause pulpal damage. Plasma membrane of the odontoblast cell bodies and odontoblast processes within the dentinal tubules are affected, and junctional complexes between odontoblasts are destroyed (Cotton WR, 1968; Avery J, 1981; Searls, J., 1975; Heys et al., 1981; Ten Cate J, 1985). Disruption of the odontoblasts results in concentration of potentiating inflammatory factors, which initiate chemotactic signals (Ohshima H, 1990). Mechanical trauma further acts by activating nerve terminals releasing neuropeptides and causing alterations in pulpal microcirculation (Kim S, 1985; Avery and Chiego, 1990). Generally, initial inflammatory tissue reactions are more extensive in the dental pulp due to chemotactic and vasoactive signals from odontoblast death and distortion of the peripheral nerve terminals. Subsequent vascular changes and inflammatory cell recruitment release histamine, seretonine and prostaglandins. These reactions are followed by the normal macrophage response, stimulating cell migration and fibroblastic activity from the adjacent fibroblasts or the perivascular cells, in the case where prolonged irritation of the pulp is prevented (Ten Cate J, 1985). Sufficient vascular function in the inflamed area and appropriate environment favor rapid and complete tissue healing. Usually, inter-odontoblastic spaces become filled with plasma leaking from the capillaries, where the clotting cascade takes place, preventing proper pulp from further ingress of irritants. Odontoblast re-establish their plasma membranes and junctional complexes and secrete extracellular matrix, progressively normalizing the whole tissue function (Avery JK, 1994). Prolonged mechanico-chemical irritation of the dental pulp and/or contamination of the injured area by oral bacteria is responsible for continuing pulp inflammation and pulp necrosis (Mjor IA, 1977; Brannstrom and Olivera, 1979).

4.3.2.2 Pulp Pathology and Dentinogenesis

A broad spectrum of formative cells and calcified matrices have been described during pathogenesis of inflammation and the subsequent pulp-dentinal repair and regeneration:

Peritubular dentin. Intratubular mineralization of the primary dentin, as a result of the aging process or in response to destructive stimuli (Fig. 4.2 B).

Reactionary tertiary dentin. Dentin which is formed at a relatively high rate post-developmentally by primary odontoblasts in response to appropriate environmental stimuli (Fig. 4.2 C), which stimulate but not affect the survival of primary odontoblasts (Smith et al., 1995).

Reparative tertiary dentin. A tubular calcifiable matrix produced by odontoblast-like cells in a polar predentin-like pattern, during the reparative process of the pulp and \ or in response to specific inductive influences, either in the odontogenic pulp periphery, or at central pulp sites. This type of matrix has also been referred to as replacement dentin, or neodentin (Tziafas D, 1995).

Odontoblast-like cells. Elongated polarized cells able to secrete tubular matrix in a polar predentin-like pattern, which are differentiated in absence of dental epithelium and/or basement membrane during the reparative process of the pulp and \ or in response to specific inductive influences. They are responsible for the production of reparative dentin either in the odontogenic pulp periphery, or in ectopic pulp sites of the pulp proper, onto physical or artificial substrata. They have also been referred to as new odontoblasts, 2nd generation odontoblasts, or replacement odontoblasts. Different phenotypic aspects have been described for odontoblast-like cells developed in various in vitro systems and in vivo experimental or clinical applications.

Primary cell cultures retain more of the phenotypic properties of pulp tissue in vivo than subcultured cells and cloned cell lines (Nakashima M, 1991; Couble et al., 2000). Odontoblast-like cells in explant cultures synthesize type I collagen (Magloire et al., 1981) and type I trimmer (Kuo et al., 1992). Pulp cells maintained in culture over a long period, exhibited formation of multi-layered structures (Mac Dougall M, 1992; Tsukamoto et al., 1992; Andrews et al., 1983), expressing type I collagen, fibronectin, vimentin, nestin and dentin phosphoprotein (Mac Dougall M, 1992; About et al., 2005). Odontoblast-like cells established from murine pulp-derived cells showed high expression of dentin phosphoprotein, type I collagen and alkaline phosphatase (Mac Dougall et al., 1995).

The differentiated cells which secrete tubular dentin during pulp repair resemble morphologically the normally developed odontoblasts. They are elongated cells with clear nuclear, cytoplasmic and secretory polarity. Limited information on the biochemical profile of biosynthetic activity of odontoblast-like cells in vivo are yet available (Magloire et al., 1988; D'Souza et al., 1995; Tziafas et al., 2000). Expression of collagen type III (Magloire et al., 1988), fibronectin (Magloire et al., 1988; Tziafas et al., 1995a; Yoshiba et al., 1996), dentin sialoprotein (D'Souza et al., 1995) and osteocalcin (Tziafas D, 1994) has been identified in reparative dentin and associated odontoblast-like cells. Nevertheless, criteria for assessment of differentiated odontoblast-like cells as used in the literature are not well defined. Cells associated with the functional activity of odontoblasts, i.e., tubular matrix deposition in a polar pattern, have been often identified as odontoblast-like cells. The staining reaction of the matrix secreted by odontoblast-like cells with antibodies raised to dentin-specific proteins would also recognize their odontoblastic specificity.

Fibrodentin. The atubular hard tissue formed by fibrodentinoblasts, either in the odontogenic pulp periphery or in ectopic pulp sites. The osteotypic form of fibrodentin is also referred to as osteodentin or trabecular dentin.

Fibrodentinoblasts. These are cuboidal or spindle-shaped pulp cells which produce atubular hard tissue. They are developed during the incomplete differentiation process of pulpal cells due to the effects of radiation or antimitotic agents, during the stereotypic process of pulp wound healing before the appearance of odontoblast-like cells, or in a non-appropriate pulp environment. Pulp cells forming osteotypic form of fibrodentin are further named osteodentinoblasts.

4.4   REGENERATION OF PULP/DENTIN-LIKE STRUCTURES: EXPERIMENTAL MODELS AND MECHANISMS

As other vertebrate organs, teeth develop from epithelial and mesenchymal tissues, sequential and reciprocal interactions between them determine cellular components of interacting tissues to respond to inductive signals in a specific way. This phenomenon is referred to as cell competence, and it is of critical importance for any tissue histogenesis (Gurdon JB, 1992). Mesenchymal cells of the dental papilla during tooth morphogenesis, progressively acquire increasing levels of specification, and as a consequence, they become the only cells which respond to signals from the enamel epithelium by differentiating into odontoblasts.

It has been repeatedly demonstrated by several experimental works that odontoblast terminal differentiation can only occur under epigenetic signals, given by stage-specific inner dental epithelium and basement membrane. On the contrary, many tissue recombinations and experimental conditions demonstrated that preodontoblasts do not differentiate into functional odontoblasts when isolated dental papillae without adjacent basement membrane and epithelial elements are cultured in vitro or implanted in vivo.

The most important experimental models used to approach expression of the dentinogenic potential of dental papilla and pulp cells are reviewed below. However, it must be emphasized that dentinogenic activities can be characterized as odontoblast differentiation and (primary) dentin formation, when preodontoblasts interact with enamel dental epithelium, under normal developmental conditions. The differentiated cells detected in the broad spectrum of ex vivo and in vivo models, leading to the so-called experimental dentinogenesis, might be referred to as odontoblast-like cells and the matrix tertiary (or reparative) dentin. This is also a fundamental issue for any study aiming toward the regeneration of the dentin-pulp complex or tooth engineering.

4.4.1   EXPERIMENTAL EX VIVO MODELS

a. Cultures of competent dental papillae cells, i.e., dividing and post-mitotic preodontoblasts, in the presence of various matrix molecules showed either maintenance of the cytological state of differentiating cells in the presence of dental pulp biomatrix (Cam et al., 1986), some non-collagenous fractions of dentinal matrix (Lesot et al., 1981), hyaluronic acid and/or chondroitin sulphate (Tziafas et al., 1988), polarization and functional differentiation of odontoblast-like cells when the papillae cultured in a semi-solid agar in the presence of various pure growth factors, such as the TGF-beta1 and BMP-2 combined with matrix molecules, heparin and fibronectin, or some dentin components (Begue-Kirn et al., 1994; Lesot et al., 1993, 1994). The new matrix contained the normal constituents of predentin in developing teeth, such as collagen type I, decorin, biglycan and fibronectin. Pre-incubation of the active dentin components with a monoclonal antibody blocking TGF-β activity led to abolishment of their biological activity. Comparing the expression patterns for TGF-βs, BMPs, IGFs, fibronectin, osteonectin, msxs and bone sialoprotein's gene between these in vitro developed odontoblast-like cells and in vivo differentiated primary odontoblasts, it was found that the same patterns exist for all molecules except msxs, which appear only in vivo. Experimental attempts have been made to give rise to odontoblast-like cell lines from dental papilla cells after their transfection with simian virus large T antigen (Mac Dougall et al., 1995) or human papillomavirus 18 (Thonemann and Schmalz, 2000), expressing dentin matrix protein-1 and the dentin sialophosphoprotein molecules (Hao et al., 2002; Kamata et al., 2004).

b. Cultures of isolated dental pulp cells. Mature animal or human dental pulp cells have been cultured in various conditions, such as directly on glass coverslips or cyanoacrylate films (Magloire et al., 1988), on collagen-chondroitin sulfate sponges (Bouvier et al., 1990), on fibronectin-coated glass (Veron et al., 1990), on microcrystals produced by the reaction of a calcium hydroxide-containing materials with culture medium (Seux et al., 1991), in the presence of hydroxy-vitamin D3 (Tsukamoto et al., 1992), or dexamethasone (Kasugai et al., 1993), in the presence of β-glycerophosphate (About et al., 2000), etc. A high number of isolated cells, alkaline phosphatase activity, partial cytodifferentiation of explanted pulp cells with modulated protein synthesis and formation of minerized nodules, were found in most of these conditions. Changes in the biosynthetic activity of cultured cells were mainly seen in the presence of the hydroxy-vitamin D3 (Tsukamoto et al., 1992), dexamethasone (Kasugai et al., 1993), crude bone morphogenetic protein from bone and dentin (Nakashima M, 1992), growth factors such as PDGF, IGF-I and II, aFGF, bFGF and TGF-β (Shiba et al., 1995; Shirakawa et al., 1994). In routine cultures, pulp cells do not express the so-called dentin-proteins (dentin phrosphoprotein and dentin sialoprotein), which have been used as dentin-specific markers for many years (D'Souza et al., 1992; McDougall et al., 1997). Despite the fact that these proteins have been recently detected also in bone cells (Qin et al., 2002), they are widely accepted as markers of odontoblast-like cell specificity in selected cultured clones of pulp cell population (Hanks et al., 1998).

c. Cultures of dental pulp progenitor/stem cells. Clonal rat pulp cells have been considered as partially differentiating odontoprogenitor cells without any morphological appearance of the odontoblastic phenotype, showing increased alkaline phosphatase activity, production of collagen type I, osteopontin synthesis and several receptor-coupled intracellular signalling systems (Liang et al., 1990; Kawase et al., 1995). Culture of porcine pulp cells, as a three dimentional pellet, promoted odontoblast differentiation compared with monolayers (Iohara et al., 2004); pulp cells were stimulated by Bone Morphogenetic Protein-2 and expressed dentin sialophosphoprotein and enamelysin (matrix metalloproteinase 20).

d. Culture of tooth organ—In this ex vivo model (Parish et al., 1995), the basic architecture of the cut tissue, the normal cell-cell and cell-matrix interactions, and the organization of microvascular system are maintained. Thick sliced human (Magloire et al., 1996) or animal teeth (Sloan and Smith, 1999) have been used to approach the mechanism underlying tertiary dentinogenesis. Implantation of heparin-agarose beads, soaked with TGF-β isoforms in odontoblast layer of rat tooth slices, showed that TGF-β3 but not TGF-β1 stimulated odontoblast-like cell differentiation (Sloan and Smith, 1999). Similarly, Magloire et al., 2001, studied the behaviour of 750μm - thick slices from human teeth cultured from 3 to 30 days. Pulp trauma developed aspects of pulp healing, such as the cell proliferation, migration, matrix formation and neovascularization; elongated spindle-shaped cells were lined along the affected area of circumpulpal dentin. Application of microtubes filled with neural growth factor/serum-free medium on dentin close to the pulp, induced after 21 days extension of cytoplasmic processes of the spindle-shaped cells within the treated dentinal tubules (D'Souza et al., 1998). In cultured human tooth slices, application of TGF-β1 containing alginate hydrogel resulted in odontoblast-like cell differentiation and reparative dentinogenesis (Dobie et al., 2002).

4.4.2   EXPERIMENTAL IN VIVO MODELS

The mechanism which controls the expression of the dentinogenic potential of dental mesenchymal cells could be only understood by using combined ex vivo and in vivo approaches, since cell or tissue culture systems cannot approach the broad spectrum of biological processes taking place during repair or regeneration of the dentin-pulp organ. Several in vivo models contributed significantly to this understanding, leading researchers to design new treatment strategies.

4.4.2.1 Non-Exposed Dentinal Cavities

Upregulation of odontoblast biosynthetic activity and odontoblast replacement has been described as a part of pulp tissue repair following local injuries in non-exposed cavities, such as the carious lesions, tooth grinding, restorative materials, prolonged air-blast, coronal tooth fractures, attrition, abrasion and erosion (Stanley and Lundy, 1972; Bergenholtz G, 1981; Trowbridge HO, 1981). Local dentinal injuries affect primary odontoblasts. Increased capillary permeability and fluid transudate have produced separate odontoblasts from the predentin; odontoblastic processes are retracted, leading to odontoblast necrosis (Heyeraas et al., 1983). Destruction of primary odontoblasts initiates chemotactic signals, resulting in accumulation of inflammatory cells (Couve E, 1986). Inflammatory cells produce the inflammatory mediators regulating pulpal blood flow and capillary permeability. The outcome of pulp response to local dentinal injuries is influenced by numerous factors; the presence of bacteria in the affected area has been repeatedly seen as the most critical factor. In the case where the insult is overcome, the pulp-dentinal complex progressively returns to pulp healing (Murray et al., 2002). Dentin permeability is dramatically reduced in the affected area, while the odontoblast injury stimulates proliferation and migration of the progenitor cells in the pulp parenchyme (Pashley DH, 1985; Kim S, 1990). The cascade of dentinogenic events has been approached at cellular and molecular levels during the last few decades:

- Production of cytokines by macrophages which further influence directly the behaviour of the underlying pulp cells or interaction with active morphoregulatory molecules released from the surrounding dentin matrix has been reported (Smith AJ, 2002; Rutherford et al., 1995).

- Biologically active molecules might provide signals that increase cell mitoses in the cell-rich zone and stimulate spindle-shaped cells to migrate towards the region of lost odontoblasts, where they become elongated polarized cells elaborating tubular matrix in a polar predentin-like pattern (Sveen and Hawes, 1968).

- The stimulated cells of the odontoblast layer form reactionary dentin while peritubular dentin formation is also seen (Murray and Smith, 2002). These events are clearly restricted to those cells affected by the injury. Peritubular dentin formation might be distinguished from the atypical intratubular calcification, which has been suggested to represent a non-vital process (Bjorndal and Darvann, 1999).

- The replacement odontoblast-like cells morphologically resemble primary odontoblasts (Yamamura T, 1985) and also synthesize some molecules detected in odontoblasts (Magloire et al., 1988). However, they cannot be considered as fully differentiated odontoblasts; early elaboration of fibrodentinal matrix very often precedes the appearance of odontoblast-like cells (Senzaki H, 1980).

- Under clinical conditions, mineralized matrix formed at the pulp-dentin interface often comprises reactionary dentin, reparative dentin and atypical fibrodentin. It is impossible to distinguish these defensive processes at the in vivo level, and the processes may also be indistinguishable from biochemical perspective (Smith AJ, 2002).

• Transdentinal stimulation of tertiary dentin formation was also seen after application of BMP-7 (Rutherford et al., 1995) or EDTA-soluble dentin matrix components (Smith et al., 2001) in unexposed cavities of monkey or ferret teeth, respectively. Recombinant TGFβ1 placed on acid-treated dentin of unexposed cavities of dog teeth (Kalyva et al., 2009) stimulated both formation of tertiary dentin, while intratubular mineralization was further observed in affected dentinal tubules.

4.4.2.2 Exposed Pulps

Numerous clinical and experimental attempts with traditional capping materials (Fig. 4.3 A) have repeatedly showed that tertiary dentin can be formed as a part of wound healing in the bateria-free pulp environment. Furthermore, several investigations have been made during last two decades to introduce biologically active materials in the treatment of pulp exposures.

• The materials which are traditionally used in pulp capping situations, calcium hydroxide-based materials, Mineral Trioxide Aggregate (MTA) and resin-based adhesive systems have been repeatedly tested for their ability to induce tertiary dentin. Numerous experimental studies in animal teeth and clinical trials generated strong support for calcium hydroxide-based materials and MTA (Bergenholtz G, 2005).

• The ability of calcium hydroxide to stimulate formation of reparative dentin and the associated cytological aspects of formative cells have been extensively studied during the last 4 decades (Fig. 4.3 B,C). Nevertheless, very few data exist in the literature concerning the molecular mechanisms of tertiary dentinogenesis in response to pulp capping with calcium hydroxide-based materials. Furthermore, the end-result of pulp capping with calcium hydroxide-based materials is not predictable. Often, a scar-like mineralized tissue is formed (Fig. 4.3 D,E).

On the other hand, the Portlant cement-based material (MTA) has only limited been investigated as capping material, but there is no evidence up to now that the mechanism by which hard tissue formation is induced is different to that by calcium hydroxide (Nair et al., 2008).

• Properly designed clinical studies using resin composites for pulp capping treatment are lacking, but it seems that tissue healing with hard tissue formation is an extremely rare response to the adhesive materials even in the case of carefully prepared mechanical exposures in healthy animal teeth.

• In general, in today's clinical practice, restoration of teeth with pulp exposures remains as one of the most problematic and unpredictable methods of treatment since 1949 (Bergenholtz G, 2005). Horsted et al., 1985 reported that success rate in post-operative periods for 5 years and more is over 80% of carefully selected treated cases. Pulpal exposure due to caries shows very limited potential for pulp survival due to bacterial infection of the pulp for a substantial

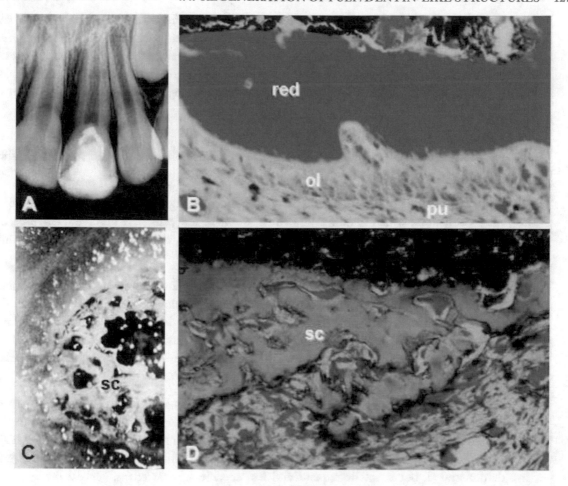

Figure 4.3: A. Human upper central incisor pulptomized and capped with MTA. Thin hard tissue bridge in contact with the capping material. B. Dog dental pulp (pu) capped with a calcium hydroxide-based material. Thick reparative dentin bridge (red) associated With a layer of odontoblast-like cells (ol). C. Dog dental pulp capped with a calcium hydroxide-based material. Formation of a scar-like hard tissue bridge (sc). D. Dog dental pulp capped with a calcium hydroxidebased material. Formation of a scar-like hard tissue bridge (osteodentin) in contact with theee capping material.

period of time, which compromises the wound healing reaction (Bergenholtz G, 2001). Case selection is a very crucial parameter in the clinical studies evaluating the outcoming of exposed pulp treatment. Thus, it is reasonable to suggest that the prognosis of pulp treatment depends on both, the existence of favourable conditions for pulp healing and the ability of a given therapeutic application to regulate pulp-dentin regeneration.

Conflicting results have been reported by numerous studies with biological materials, such as the cortical steroids, antibiotics, extracellular matrix molecules, tricalcium phosphate ceramics, biomatrices and growth factors/bone morphogenenetic proteins. Since the last 4 groups of materials display inductive properties for the various processes taking place during the wound healing process, they have been tested experimentally for many tissue engineering applications in the dentin-pulp complex.

- Collagens – Rapid pulp healing by reduced ability to form complete dentinal barriers were found in response to application of collagen sponge or enriched collagen solution in direct pulp capping situations (Dick and Carmichael, 1980; Fuks et al., 1984).

- Hyaluronic acid – In hyaluronic acid-treated pulps, healing events showing most of the cytological features of newly diffferentiated odontoblast-like cells were observed; however, this acromolecule seems to influence the environment of the pulp rather than being stimulatory (Sasaki and Kawamata-Kido, 1995).

- Dentinal constituents – Application of a dentin extracellular matrix compenent, bone sialoprotein with gelatin as a carrier resulted in extensive formation of atubural fibrodentin within the pulp chamber (Decup et al., 2000). In the same model, dentonin, a peptide able to stimulate proliferation of dental pulp stem cells and their differentiation in vitro (Liu et al., 2004), induced mineralization that transformed the central pulp into a homegenous mineralized tissue (Goldberg et al., 2006).

- Tricalcium phosphate ceramics – Despite the fact that hydroxyapatite has shown clear osteoconductive and occasionally osteoinductive potential, controversial results have been reported concerning its effect as pulp capping material (Yoshida et al., 1992; Jean et al., 1993; Holtgrave and Donath, 1995). Putative dentinogenic ability of these materials has been demonstrated, but it was infrequent and less specific (slower effect, general tissue calcification) compared to calcium hydroxide (Furusawa et al., 1991; Yoshimine and Maeda, 1995). The slightly alkaline calcium-β-glycerophosphate (Imai and Hayashi, 1993) and the bioactive ceramic bioglass (Oguntebi et al., 1988) induced formation of tubular dentin in a more specific manner.

- Biomatrices – Since reparative dentin formation has been consistently found in close proximity to dentinal chips unintentionally penetrating sites of proper pulp during mechanical pulp exposures (Stanley HR, 1989), many investigators tried to induce dentinal bridging of exposed pulps by using demineralized dentin (Anneroth and Bang, 1972; Nakashima M, 1989). The sequence of cellular events in dog pulps capped with antigen-extracted allogenic dentin was described by Nakashima M, 1989: Spindle-shaped and large mesenchymal cells migrating towards the amputated tissue attach to the used biomatrix, while capillary infiltration is also seen. Attached spindle-shaped cells formed osteotypic fibrodentin, which preceded any further differentiation of the large cells into elongated polarized cells elaborating reparative dentin. Guanidine-inactivated dentinal matrix (after removal of growth factors) in the same experiment demonstrated also inductive activity, but it was delayed, indicating that the used matrix as capping agent may provide a suitable scaffolding for attachment of pulp cells. Further differentiating events controlled by endogenous factors have been hypothetized.

- Growth factors and bone morphogenetic proteins – Biologically active molecules such as enamel matrix derivative gel, the BMP-2, -4 or –7 or (Lanjia et al., 1993; Rutherford et al., 1993; Suwa et al., 1993; Nakashima et al., 1994; Nakashima M, 1994; Six et al., 2000), EDTA-soluble dentin components (Smith et al., 1990), and TGF-β1 (Nakashima et al., 1994) were also placed in contact with amputated pulp in capping situations. The stimulatory effect of these molecules (except the TGF-β1) on the pulp repair was clear in all studies; enhancement of morphogenetic events in the repairing pulp, i.e., formation of fibrodentin matrix followed by reparative dentin formation was often found. Tubular reparative dentin with odontoblast-like cells was found after application of dentin matrix proteins in young ferrets; osteodentin deposition was seen in older animals (Smith et al., 1990). More hard tissue was found after application of osteogenic protein -1 than in Ca(OH)2-treated pulps (Rutherford et al., 1993). Recombinant BMP delivered in a scaffold of demineralized dentin matrix induced tubular dentin, whereas the same signaling molecule delivered in collagen type I matrix induces osteodentin (Nakashima M, 1994). The reparative dentin formed in the above mentioned capping studies appears, initially, as a zone of osteodentin matrix with many soft tissue and cellular inclusions, which progressively changes into dentin-like matrix associated with elongated polarized cells forming tubular matrix in a polar predentin-like pattern.

- Cell or gene-based biological applications – Skin fibroblasts transduced with osteogenic Protein-1-adenovirus induced reparative dentin formation in ferret dental pulps (Rutherford RB, 2001). Pulpal stem cells cultured as a 3-D pellet and stimulated by the BMP-2 induced reparative dentin after their implantation in pulpal sites (Iohara et al., 2004). Ultrasound gene delivery of the growth/differentiation factor 11 stimulated reparative dentin formation in sites of pulp capping (Nakashima et al., 2003). These experimental attempts have been recognized as useful tools for cell–based tissue engineering therapies. However, since pulpal cells cultured in the presence of specific scaffolds and implanted in vivo are able to express numerous biologically active molecules (Burma et al., 1999), the exact biological mechanism underlying these inductive effects cannot be easily studied in the cell- or gene-based approaches.

The placement of biologically active molecules in capping situations represents a complex reparative process because of inflammatory reactions, mechanical trauma of exposure, bleeding during surgery, possible release of growth factors from circumpulpal dentin, etc. (Lesot et al., 1994; Tziafas et al., 2000). Nakashima and Reddi, 2003 emphasized the role of extracellular matrix scaffolds in determining the differentiation of odontoblast-like cells in such capping experiments by using biologically active molecules. However, the role of growth factors involved in the effects of bioactive applications remains unclear; since the pulp capping model does not offer a very suitable model to study the tissue-specific mechanisms of direct biological effects of applied molecules on pulp cells due to the dynamic process of wound healing in the treated area (Rutherford et al., 1999; Tziafas et al., 2000), other experimental approaches have been used to clarify the dentinogenic potential of pulpal cells: implantation of biologically active materials at central pulp situation.

Pulp cell responses to implantation of biomatrices or biomolecules into central pulp, at a distance from the site of mechanical pulp exposure, have been analysed in a series of experiments by our research team. Obviously, the normal sequence of wound healing events do not take place in this model, but the direct effects of exogenous influences on pulp cells with a minimal tissue trauma could be approached (Fig. 4.4 A). The data are focused on interactions between pulp cells and exogenous influences in acquisition of the odontoblast-like cell phenotype and initiation of reparative dentin formation.

The data obtained by this model are presented in brief below:

- The response of dental pulp and papilla ectomesenchymal cells to demineralised with acetic acid dentin and bone matrices was evaluated after periods of 2 or 3 weeks (Tziafas and Kolokuris, 1990). The demineralized or native dentin implants exhibited a promoting effect on differentiation of ectomesenchymal cells of young pulp (Fig. 4.4 B) and papilla sites into matrix forming cells (Tziafas et al., 1992). A characteristic difference in the calcifying potential of dental pulp and papilla tissues, which consist of cell population of common origin, was seen. The demineralized bone implants exhibited a promoting effect only in pulp sites. The interactions between pulp cells of adult mature teeth and autogenous dentin matrix were further investigated in order to compare the expression pattern of odontoblast-like cell phenotype and initiation of reparative dentinogenesis in adult mature teeth with that described previously in young teeth (Tziafas et al., 1993). Two mechanisms for induction of dentinogenesis were described: direct induction of differentiating odontoblast-like cells and indirect matrix synthesis, leading to cytodifferentiation of odontoblast-like cells (Tziafas D, 1997). Physico-chemical properties of a surface to which pulp cells attach are a critical requirement for modulation of cell activity. Predentin is the most appropriate surface for adhesion of pulp cells. Demineralized dentine showed co-existence of adjacent polarizing and non-polarized cells along the implants. The organic framework in native dentin is strongly masked by the inorganic salts, delaying the modulation of interacting cells' activities. The mature pulp cells maintain the ability to differentiate into dentin-forming cells as a direct response to exogenous inductive influence. Implantation of native dentinal matrix (mineralized or predentin) exhibited a clear dentinogenic activity on mature pulp cells, but demineralized dentin did not. Despite the clear dentino-inductive properties of the active molecules, which are present in the EDTA-soluble fraction of dentin (Begue-Kirn et al., 1994), their role have not been considered as essential for the expression of the dentinogenic activity of the pulp-dentin complex.

- Immuno-electron and immuno-fluoresence analyses of the involvement of fibronectin in dentin-induced dentinogenesis were performed (Tziafas et al., 1995a). The responses of den-

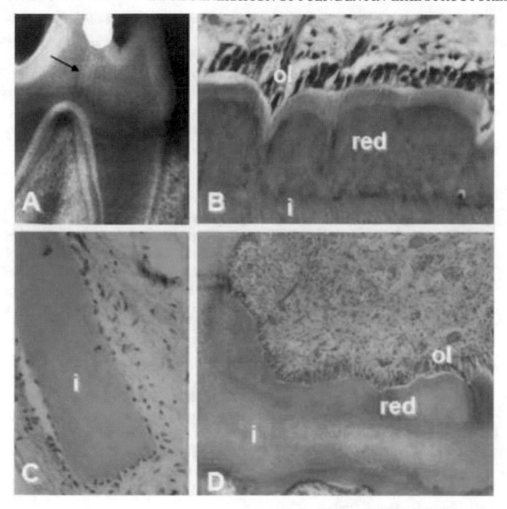

Figure 4.4: A. Dog molar immature tooth. Implantation of demineralized dentin (arrow) at the central pulp. B. Formation of reparative dentin in direct contact with demineralized dentin implant (i). Newly differentated odontoblast-like cells (ol). C. Absence of any tissue organization or matrix formation around the millipore filter implant (i). D. Formation of reparative dentin associated with odontoblast-like cells in direct contact with millipore filter implant (i) that had been soaked in solution containing TGF-beta1.

tal pulp cells to intrapulpal implantation of Millipore filters, containing exogenous plasma fibronectin or a fibronectin-like protein polymer, were evaluated (Tziafas et al., 1992b; Tziafas D, 1995). Immunoelectron microscopy of demineralized dentine implants that had been in the pulp for 3 days showed heavy adsorption of fibronectin on to their surfaces. Differentiating odontoblast-like cells synthesize fibronectin, as do developing preodontoblasts, but this biosynthetic function seems to be retained in mature odontoblast-like cells. Exposure of ectomesenchymal cells of dental pulp parenchyma to Millipore filters, immersed in a high concentration of plasma fibronectin, showed a layer of cells exhibiting odontoblast-like features (size, shape, and polarity) and tubular matrix formation. Implantation of filters containing the fibronectin-like polymer did not show any evidence of cellular organization or matrix synthesis. The synthetic peptide used with many RGD, sequenced from human fibronectin, did not induce any morphogenetic event (cellular organization or matrix synthesis). Millipore filters coated with fibronectin failed to induce any cellular organization or matrix synthesis, even after 10 weeks of their exposure to the adult pulp cells.

- The effects of dentin extracellular matrix components on dental mesenchymal cells were studied (Tziafas et al., 1995b) by light and transmission electron microscopy, 8 and 24 days after their implantation. The role of TGF-β as an active component of the dentin matrix during induction of reparative dentinogenesis was studied (Tziafas and Papadimitriou, 1998). The dentinogenic activity of dentin matrix may, at least partly, be ascribed to TGF-β molecule(s). The activity of demineralized dentin matrix after implantation in sites of central dog dental pulp parenchyma was abolished by pre-incubation with an antibody blocking the activity of TGF-β.

- Recombinant bFGF , IGF-II and TGF-β1 after their implantation at central pulp sites in vivo demonstrated regulatory effects on pulp cells (Tziafas et al., 1998). bFGF stimulated formation of osteotypic extracellular matrix without any evidence of reparative dentin-specific activity. IGF-II did not show any particular response; only stimulation of osteotypic matrix formation at a distance from the implants was seen. TGF-β1, both as purified or recombined molecule, specifically induced cytological and functional differentiation of odontoblast-like cells and tubular dentin-like matrix after implantation (Fig. 4.4 C and D).

4.4.2.3 Transplantation of Pulpal Cells

Many sites have been used for implantation of isolated pulp tissue, such as subcutaneous connective tissue (Zussman WV, 1966; Inoue and Shimono, 1992), brain (Luostarinen and Ronning, 1977), muscular tissue (Takei et al., 1987), kidney (Yamamura T, 1985; Braut et al., 2003), bone tissue (Takei et al., 1987), spleen (Ishizeki et al., 1989), etc. In general, we find the following:

- Grafted odontoblasts from adult rat pulp degenerated but osteodentin matrix had been formed by newly differentiated cells (Zussman WV, 1966).

- Pulp cells enclosed by intact odontoblast layer, which had been dissociated from the surrounding dentin by treatment with EDTA, formed tubular dentin by the peripheral odontoblasts and osteodentin by other pulpal or papilla cells when transplanted into spleen (Ishizeki et al., 1989; Braut et al., 2003).

- Dissociation of dental pulp organ from the surrounding structures and their re-association after the chemical treatment of circumpulpal dentinal walls with citric acid, hyaluoridase, sodium hypochloride and oxygen peroxide showed differentiation of new generation odontoblasts after two weeks implantation into connective tissue (Heritier et al., 1989).

- Implantation of alone pulp tissue or in association with the pulp chamber resulted in the same results as in subcutaneous grafts; implantation of these tissues within diffusion chambers did not show any dentin or osteodentin formation (Takei et al., 1987). Yamamura T (1985) demonstrated that autografting of dental pulp under the kidney capsule or into the anterior chamber of the eye, hard-tissue forming cells are developed from proper pulp following the degeneration of primary odontoblasts.

- Grafting of either pulp tissue or pulp tissue within the pulp chamber into medullary cavity or extraction socket showed only traces of new dentin adjacent to the pulp chamber wall. Recently, Braut et al., 2003 used coronal pulp tissue isolated form postnatal marked transgenic animals and transplanted them under the kidney capsule; they provided for the first time direct evidence that the dental pulp contains both, progenitors able to differentiate into cells forming reparative dentin with distinct dentin sialophosphoprotein expression and cells forming bone-like tissue with dentin matrix protein-1 expression.

- Recombinants of murine epithelial root sheath and dental mesenchyme resulted in differentiation of odontoblast-like cells and fragments of dentin (Thomas and Kollar, 1989).

- Findings from all these experiments have clearly shown that removal of pulp tissue from its specific micro-environment lead pulp cells to differentiate into osteodentin matrix formation. Odontoblasts or cells elaborating any tubular matrix were never seen in grafts of pulp alone. On the other hand, transplantation of pulp tissue or dental pulp cells within its surrounding tooth structure can occasionally lead to appearance of dentin-like structures.

4.5   DESIGN PRINCIPLES IN THE REGENERATIVE TREATMENT STRATEGIES OF PULP/DENTIN COMPLEX

4.5.1   THE RESTORATIVE VS REGENERATIVE CONCEPT IN OPERATIVE DENTISTRY

Researchers working in life sciences are increasingly using tissue engineering-based approaches to treat a variety of pathological conditions. Tissue engineering is a multi-disciplinary field which belongs to the so-called regenerative medicine that brings together basic biological sciences, engineering and clinical sciences aiming to develop new tissues and organs. The ultimate goal of regenerative treatment strategy is to reconstitute lost tissues and to improve altered tissue functions.

For many years, dental practitioners have been using regenerative approaches to stimulate repair of tooth structure affected by caries, physical trauma or iatrogenic procedures. Over the last decade, dental scientists started to explore the potential of new regenerative and tissue engineering-based strategies to design new therapies in clinical dentistry, to replace dental tissue structures and to maintain or improve dental tissue functions. In general terms, the concept of regenerative treatment strategies is to move from the traditional aim of operative dentistry to replace lost tooth structure with the appropriate dental materials, to the new one of modern biomedical sciences to replace lost tooth structure with tooth structure.

The treatment of dental cavities with and without pulp exposure, the exploitation of the regenerative capacity for the reconstitution of lost tooth structure in carious or traumatized teeth, and the continuation of root development in immature non-vital teeth are only indicative examples of the new directions of research in the field of pulp/dentin complex biology and regeneration.

4.5.2   IMPROVING DENTINAL REACTIONS IN NON-EXPOSED CAVITIES (NON-EXPOSED DENTINAL CAVITY SITUATION)

For successful long-term outcome of any operative dental treatment performed on vital teeth without pulp exposure, a prime aim should be to stimulate pulp/dentin responses (Fig. 4.5 A), leading to a tissue able to effectively oppose exogenous destructive stimuli including bacteria, toxins, and toxic substances from materials to protect the pulp. Thus, the ultimate goal of a regenerative treatment strategy in the dentinal cavities is to reduce dentin permeability, as occurs physiologically in dentin in response to slowly progressive carious lesion (biomimetical approach).

It has been recognized from the sixties (Cotton WR, 1968; Sveen and Hawes, 1968) that the transdentinal reactions to the so-called "operative trauma," including physical and chemical effects, depend on the extent of injury to the underlying pulp and the treatment modalities, such as the operative procedures to remove the injurious challenges, the biological effects of the applied materials and the control of post-operative infection (Tziafas et al., 2000).

In cases of mild dentinal injuries, where odontoblasts may survive, e.g., slowly progressing dentinal caries, mechanico-chemical irritation or fracture in cavities with remaining dentin thickness more than 0,5 mm, abrasion and erosion (Tziafas D, 2004), the optimum end result should be

(i) short-term stimulation of peritubular dentin formation in affected dentinal tubules (dentinal sclerosis) and

(ii) a regional and time limited up-regulation of the biosynthetic activity of survived odontoblasts (reactionary dentin formation).

On the other hand, in the case of severe injury of the dentin-pulp complex, where odontoblasts are destroyed subjacent to the dentinal cavity, e.g., rapidly progressing dentinal caries, severe pulpal injury due to mechanico-chemical injury or fracture in cavities with remaining dentin thickness less than 0,3 mm, the regenerative treatment might has as an optimum end result the replacement of lost odontoblasts by newly differentiated odontoblast-like cells and a time-limited formation of reparative dentin (Smith et al., 1995). Early formation of atubular fibrodentin (or interface dentin according to Mjor IA, 1983) just after the destruction of odontoblasts should provide the necessary barrier effect to decrease the permeability of the residual primary dentin.

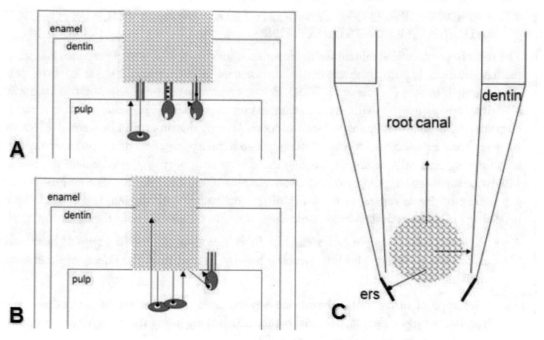

Figure 4.5: Diagrammatic representations of the working hypotheses on the mechanisms by which new treatment strategies (t-s) regenerate the dentin/pulp complex. A. Non-exposed dentinal cavity situation. Bioactive molecule(s)-based application, stimulating transdentinally the survived primary odontoblasts to form peritubular and reactionary tertiary dentin, and/or pulpal cells to migrate towards the circumpulpal dentin, differentiate into odontoblast-like cells and form reparative tertiary dentin. B. Pulp capping situation. Bioactive molecule(s)-based application, inducing pulpal cells to migrate towards the surface of capping material, differentiate into odontoblast-like cells and form reparative tertiary dentin, or to invade the lost dentin space and form hard tissue progressively replacing the applied matrix. C. Pulpless immature tooth apex situation. Dental pulp stem cells-based tissue engineering leading to growth of dental pulp, migration of pulp cells towards the circumpulpal dentin, differentiation into odontoblast-like cells and formation of reparative dentin, and interaction with the epithelial root sheath (ers) continuating root development.

Whilst development of systems for delivery of exogenous signaling molecules has been investigated during the last 15 years (Rutherford et al., 1995; Smith et al., 2001; Kalyva et al., 2009) and seems to be a significant task for the future, the endogenous pools of biologically active molecules in the dentin matrix may provide an alternative pathway of treatment (Smith AJ, 2002). The advantages of the latter strategy is that it is a natural delivery system, and their biological activity could be protected since they are associated with the dentin matrix components until released.

4.5.3   DEVELOPING DENTIN-LIKE STRUCTURES IN PULP EXPOSURES (PULP CAPPING SITUATION)

The most important determinant for the maintenance of pulp vitality and function, following treatment of the exposed pulp, is the restoration of normal tissue architecture (Fig. 4.5 B) at the amputated dentin-pulp interface (Tziafas et al., 2001). Primary dentin is the most appropriate tissue which can effectively oppose exogenous destructive stimuli. However, since primary odontoblasts at the exposed pulp are destroyed and they are post-mitotic cells, primary dentin cannot be formed. Thus, the ultimate goal of a regenerative treatment strategy is to induce generation of a layer of odontoblast-like cells forming reparative dentin at the amputated pulp periphery and to stimulate the biosynthetic activity of surrounding primary odontoblasts forming reactionary dentin. The end result should be a complete bridge of reparative dentin in direct continuum with surrounding reactionary dentin, providing optimal physical and mechanical properties to the tooth under the following conditions:

- Reparative dentin might be formed as a direct biological effect of the applied biomaterial and not as a part of tissue healing, which is usually delayed from the effects of other exogenous influences.

- This type of bridge can only provide a predictable barrier effect to protect the pulp from secondary pulp infection, bacterial byproducts leaking along the margins of restorations and toxic substances from restorative materials.

- The specific nature of the mechanism leading to direct induction of reparative dentinogenesis lead to a time-limited formation of calcified tissue at the pulp periphery and not at the expense of underlying pulp, avoiding uncontrolled obliteration of the pulp chamber.

Furthermore, in cases with extensive destruction of coronal dental tissues due to caries or physical trauma, where dentists do not have an ideal solution for the patient's needs (Nor, 2006), the underlying non-exposed pulp should offer its regeneration potential to strengthening the whole tooth structure. The end result of a such regenerative strategy should be the partial, at least, replacement of lost dentin with new hard tissue (Jepsen et al., 1997). Rutherford et al. (1993) showed that the size and shape of the applied biomaterial on exposed pulp controls the size and shape of the new calcified tissue. Depending on the treatment modality, mainly on the type of extracellular matrix used as a scaffold, osteodentin can be formed at the site of lost dentin, restoring the coronal dental tissues. Progressively more tubular dentin associated with odontoblast-like cells might be formed providing the critical requirement for long term pulp survival.

4.5.4   ENGINEERING DENTAL PULP IN IMMATURE TEETH (PULPLESS IMMATURE TOOTH APEX SITUATION)

Pulp inflammation and/or necrosis and apical periodontitis, interrupt the development of the incompletely formed roots in developing teeth. The presence of an open apex and very thin and fragile dentinal walls presents a problem which complicate the clinical management of pulp and periapical disease. Numerous therapies have been recommended to induce continuation of the root-tip formation (apexogenesis) or apical closure (apexification). Apexification approaches by using calcium hydroxide (Felippe et al., 2006), tricalcium phosphate (Harbert H, 1996), MTA (Hachmeister et al., 2002; Felippe et al., 2006), dentin powder, and biologically active molecules or matrices (Thibodeau et al., 2007) reported apical calcified tissue formation bridging the open apex. New dentin formation and differentiation of odontoblast-like cells at the apical bridge or along the dentinal walls into the root canal space have never been reported. In a few studies, deposition of tubular dentin was attributed to the presence of remnants of vital pulp (Torneck et al., 1973a,b; Yoshida et al., 1998).

Recent studies addressed tissue engineering strategies aiming to reconstitution of a functional pulp/dentin organ at the root apex and continuation of root development (Fig. 4.5 C). These studies are based on the potential role of progenitor/stem cells, which may participate in dental regeneration (Goldberg et al., 2004). Mesenchymal stem cells isolated from dental pulp or apical papilla, the tissue which underlies dental pulp beyond the open apex have been proposed to be used for application in developing teeth with necrotic pulp. Stem cells isolated from apical papilla of the developing tooth apex exhibited, in cultures, greater numbers of population doublings, numbers of STRO-1 positive cells and tissue regeneration capacity when compared with dental pulp stem cells (Huang et al., 2008), while after their implantation into immuno-compromised mice, they generated dentin-like structure on the surface of hydroxyapatite/tricalcium phosphate carrier (Sonoyama et al., 2006). It seems that isolated dental pulp stem cells can be directed to differentiate into cells of odontoblastic lineage forming the dentin-like structure and the associated pulp tissue in transplants (Young et al., 2002; Prescott et al., 2008; Zhang et al., 2009), leading researchers to design new treatment modalities for the necrotic immature teeth. The ultimate goal should be the progressive regeneration of dentin-pulp complex, at least at the apical half of the root, by stem cells-engineering approaches, in order to strengthen the root dentinal walls and to give rise to the epithelial root sheath to continue root development.

However, a number of aspects related to the nature of dental pulp or apical papilla cell populations, as well as the characterization and determination of their potentialities in terms of specificity for regeneration of the dentin-pulp complex (Sloan et al., 2007) are still open for discussion, while the capacity of the apical papilla stem cells to differentiate into odontoblast-like cells need to be investigated further. In general, terms, it is not clear if the dentin-like forming cells in transplants could be safely characterized as odontoblasts, i.e., elongated polarized cells able to form the complete repertoire of dentinal components in a polar predentin-like pattern. In any case, data show that the problem of necrotic immature teeth in clinical endodontology will be solved by cell-based therapies and tissue engineering strategies in the near future.

BIBLIOGRAPHY

About I, Camps J, Burger AS, Mitsiadis TA, Butler WT, Franquin JC. Polymerized bonding agents and the differentiation in vitro of human pulp cells into odontoblast-like cells. Dent Mater (2005);21:156–63. DOI: 10.1016/j.dental.2004.02.011 121

About I, Bottero MJ, de Denato P, Camps J, Franquin JC, Mitsiadis TA. Human dentin production in vitro. Exp Cell Res. 2000: 258: 33–41. 123

Agata H, Kagami H, Watanabe N, Ueda M. Effect of ischemic culture conditions on the survival and differentiation of porcine dental pulp-derived cells. Differentiation (2008);76:981–93. DOI: 10.1111/j.1432-0436.2008.00282.x

Akai M. and Wakisaka S. (1990). Distribution of peptidergic nerves. In: Inoui R., Kudo T. Olgart L.(eds). Dynamic aspect of dental pulp. p.p. 337–348.Champan and Hall, London. 119

Amar S, Sires B, Sabsay B, Clohisy J, Veis A. The isolation and partial characterization of a rat incisor dentin matrix polypeptide with in vitro chondrogenic activity. (1991) *J Biol Chem*. 1991: 266: 8609–8618. 115

Anneroth G, Bang G. The effect of allogeneic demineralized dentin as a pulp capping agent in Java monkeys. (1972) *Odontol Revy*. 1972: 23: 315–328. 128

Andreasen J.O.(1970). Etiology and pathogenesis of traumatic dental injuries. Scand. J. Dent. Res., 78:329. DOI: 10.1111/j.1600-0722.1970.tb02080.x 120

Andreasen, J.O. and Andreasen, F.M.(1994):Textbook and Color Atlas of Traumatic Injuries to the Teeth.3rd Ed. Copenhagen and St. Louis, Munksgaard and C.V. Mosby. 120

Andrews CH, England MC Jr, Kemp WB. Sickle cell anemia: an etiological factor in pulpal necrosis. (1983) *J Endod*. 1983: 9: 249–252.. DOI: 10.1016/S0099-2399(86)80023-1 121

Avery J. Repair potential of the pulp. (1981) *J Endod*. 1981: 7: 205–212. DOI: 10.1016/S0099-2399(81)80177-X 120

Avery JK. Oral development and histology. 2nd edition. Thieme Medical Publishers, Inc.New York, 1994. 121

Avery, JK, Chiego DJ (1990): Cholinergic stystem and the dental pulp. In:Inoki R, Kudo T, Olgart LM, eds, Dynamic Aspects of the Dental Pulp-Molecular Biology, Pharmacology and Pathophysiology New York: 297–332. 120

Batouli S, Miura M, Brahim J, Tsutsui TW, Fisher LW, Gronthos S, et al. Comparison of stem-cell-mediated osteogenesis and dentinogenesis. J Dent Res (2003);82:976–81.

Baume LJ. The biology of pulp and dentine. In Myers HM (ed): (1980) *Monographs in Oral Science*. Basel, Karger, 1980: 8: 67–182. 115, 117, 119, 120

Begue-Kirn C, Smith AJ, Loriot M, Kupferle C, Ruch JV, Lesot H. Comparative analysis of TGF beta s, BMPs, IGF1, msxs, fibronectin, osteonectin and bone sialoprotein gene expression during normal and in vitro-induced odontoblast differentiation. *Int J Dev Biol*. 1994, 38:405–420. 123, 130

Bergenholtz G. Inflammatory response of the dental pulp to bacterial irritation. (1981) *J Endod.* 1981: 7: 100–104. DOI: 10.1016/S0099-2399(81)80122-7 120, 125

Bergenholtz, G. (1977): Effect of bacterial products on in flammatory reactions in the dental pulp. Scand. J. Dent. Res., 85:122 DOI: 10.1111/j.1600-0722.1977.tb00542.x 120

Bergenholtz G. Factors in pulpal repair after oral exposure. (2001) *Adv Dent Res.* 2001:15: 84. DOI: 10.1177/08959374010150012201 127

Bergenholtz G. Advances since the paper by Zander and Glass (1949) on the pursuit of healing methods for pulpal exposures: historical perspectives. *Oral Surg Oral Med Oral Pathol Oral Radiol Endod.* 2005: 100 (Suppl 2): 102–108. DOI: 10.1016/j.tripleo.2005.03.032 126

Bernfield, M., Kokenyesi, R., Kato, M., Hinkes, M., Spring, J., Gallo, R. and Lose, E.(1992):Biology of the syndecans. Annu. Rev. Cell Biol. 8:333. DOI: 10.1146/annurev.cb.08.110192.002053 116

Bessho K, Tanaka N, Matsumoto J, Tagawa T, Murata M. Human dentin-matrix-derived bone morphogenetic protein. (1991) *J Dent Res.* 1991: 70: 171–175. 115, 118

Bishop MA. Evidence for tight junctions between odontoblasts in the rat incisor. (1985) *Cell Tissue Res.* 1985: 239:137–140. DOI: 10.1007/BF00214913 119

Bjorndal L, Darvann T. A light microscopic study of odontoblastic and non-odontoblastic cells involved in tertiary dentinogenesis in well-defined cavitated carious lesions. (1999) *Caries Res.* 1999: 33: 50–60. DOI: 10.1159/000016495 125

Bouvier M, Joffre A, Magloire H. In vitro mineralization of a three-dimensional collagen matrix by human dental pulp in the presence of chondroitin suphate. Arch. Oral Biol. 1990: 35: 301. 123

Brannstrom, M.(1961): Cavity preparation and the pulp. Dent. Prog., 2:4. 120

Brannstrom, M.(1963):Dentinal and pulpal response,III. Application of an air stream to exposed dentin. Long observation period. In Sensory Mechanisms in Dentine Edited by D.J. Anderson. New York, Macmillan Co., pp.235–252 120

Brannstrom, M., and Lind, P.O (1965): Pulpal response to early caries. J. Dent. Res., 44:1045. 120

Brannstrom, M., and Olivera, V.(1979):Bacteria and pulpal reactions under silicate cement restorations. J. Prosthet. Dent., 41:290 120, 121

Braut A, Kollar EJ, Mina M. Analysis of the odontogenic and osteogenic potentials of dental pulp in vivo using a Col1a1–2.3-GFP transgene. (2003) *Int J Dev Biol,* 2003:47:281-92. 132, 133

Bronckers, A.L.J.J., Lyaruu , D.M. and Woltgens, J.H.M. (1989):Immunocytochemistry of extracellular matrix proteins during various stages of dentinogenesis. Connect. Tissue Res.22:65. 114, 115

Burma B, Gu K, Rutherford RB. Transplantation of human pulpal and gingival fibroblasts attached to synthetic scaffolds. Eur J Oral Sci, (1999):107:282–9.
DOI: 10.1046/j.0909-8836.1999.eos107408.x 129

Butler, W.T. (1972):The structure of a1-CB3 , a cyanogen bromide fragment from the central portion of the a1 chain of rat collagen. The tryptic pertides from skin and dentin collagen. Biochem, Biophys. Res. Commun. 48:1540. 115

Butler, W.T. (1984):Matrix macromolecules of bone and dentin Coll. Relat. Res. 4:297. 115

Butler, W.T., Bhown, M., Dimuzio, M.T. and Linde A. (1981): Noncollagenous proteins of dentin. Isolation and partial characterization of rat dentin proteins and proteoglycans using a three-step preparative method. Coll. Relat. Res. 1:187. 115

Butler, W.T., Bhown, M., Dimuzio, M.T., Cothran, W.C. and Linde A.(1983):Multiple forms of rat dentin phosphopreteins. Arch. Biochem. Biophys. 225:178.
DOI: 10.1016/0003-9861(83)90021-8 115

Butler, W.T., Bhown M., Brunn, J.C. D;souza, R.N., Farach-Carson M.C., Happonen, R-P., Schrohenloher, R.E., Seyer, J.M., Somerman, M.J., Foster, R.A., Tomana, M. and Van Dijk, S. (1992): Isolation , characterization and immunolocalization of a 53-kDa dentin sialoprotein (DSP). Matrix 12:343. 115, 118

Butler, W.T. and Ritchie, H. (1995). The nature and functional significance of dentin extracellular matrix proteins. Int. J. Dev. Biol. 39:169.. 114

Butler WT, Brunn JC, Qin C. Dentin extracellular matrix (ECM) proteins: comparison to bone ECM and contribution to dynamics of dentinogenesis. Connect Tissue Res 2003;44 Suppl 1:171–8. DOI: 10.1080/713713626

Byers M.R., Wheeller E.F. and Bothwell M. (1992). Altered expression of NGF and p75NGF-receptor by fibroblasts of injured teeth precedes sensory nerves sprouting. Growth factors 6:41.
DOI: 10.3109/08977199209008870 118

Byers MR, Schatteman GC, Bothwell M. Multiple functions for NGF receptor in developing, aging and injured rat teeth are suggested by epithelial, mesenchymal and neural immunoreactivity. (1990) *Development*. 1990: 109: 461–471. 118

Cam Y., Meyer, J.M., Staubli, A. Ruch I.V. (1986): Epithelial - mesenchymal interactions:effects of a dental biomatrix on odontoblasts. Archives of Anatomy, Microbiology and Morphology Experiments 75:75. 123

Cam Y, Neumann MR, Ruch JV. Immunolocalization of transforming growth factor beta 1 and epidermal growth factor receptor epitopes in mouse incisors and molars with a demonstration of in vitro production of transforming activity. (1990) *Arch Oral Biol*. 1990: 35:813–822.
DOI: 10.1016/0003-9969(90)90006-V 117

Cassidy N, Fahey M, Prime SS, Smith AJ. Comparative analysis of transforming growth factor –beta (TGF-β) isoforms 1–3 in human and rabbit dentine matrices. Arch. Oral Biol. 1997:42:219–223. 117

Chiba M, Nakagawa K, Mimura T. DNA synthesis and cell division cycle at the base of the maxillary incisor tooth of the young rat. Arch of Oral Biol. 1967: 12: 865–72. 114, 117

Cotton WR. Pulp response to cavity preparation as studied by the method of thymidine 3H autoradiography. In (1968) *Biology of the Dental Pulp Organ* (Ed. S.B. Finn). University of Alabama Press, Tuscaloosa, 1968:60:102. 120, 134

Cox CF, Bergenholtz G, Fitzgerald M, Heys DR, Heys RJ, Avery JK, Baker JA. Capping of the dental pulp mechanically exposed to the oral microflora -a 5 week observation of wound healing in the monkey. (1982) *J Oral Pathol.* 1982: 11: 327–339. 120

Couble ML, Farges JC, Bleicher F, Perrat-Mabillon B, Boudeulle M, Magloire H. Odontoblast differentiation of human dental pulp cells in explant cultures. Calc. Tissue Int. (2000):66:129–138 DOI: 10.1007/PL00005833 121

Couve E. Ultrastructural changes during the life cycle of human odontoblasts. (1986) *Arch Oral Biol.* 1986: 31: 643–651. DOI: 10.1016/0003-9969(86)90093-2 120, 125

Decup F, Six N, Palmier B, Buch D, Lasfargues JJ, Salih E, Goldberg M: Bone sialoprotein-induced reparative dentinogenesis in the pulp of rat's molar (2000) . *Clin Oral Invest* 2000;4:110–119. DOI: 10.1007/s007840050126 128

Dick HM, Carmichael DJ (1980). Reconstituted antigen-poor collagen preparations as potential pulp-capping agents. *J Endod.* 1980: 6: 641–644 DOI: 10.1016/S0099-2399(80)80165-8 128

Dobie K, Smith G, Sloan AJ, Smith AJ. Effects of alginate hydrogels and TGF-beta 1 on human dental pulp repair in vitro. (2002) *Connect Tissue Res.* 2002: 43: 387–390. DOI: 10.1080/03008200290000574 124

D'Souza RN, Happonen RP, Ritter NM, Butler WT. Temporal and spatial patterns of transforming growth factor-beta 1 expression in developing rat molars. (1990) *Arch Oral Biol.* 1990: 35: 957–965. DOI: 10.1016/0003-9969(90)90015-3 117

D'Souza RN, Bronckers AL, Happonen RP, Doga DA, Farach-Carson MC, Butler WT. Developmental expression of a 53 KD dentin sialoprotein in rat tooth organs. (1992) *J Histochem Cytochem.* 1992: 40: 359–366 114, 124

D'Souza RN, Bachman T, Baumgardner KR, Butler WT, Litz M. Characterization of cellular responses involved in reparative dentinogenesis in rat molars. (1995) *J Dent Res.* 1995: 74: 702–709. DOI: 10.1177/00220345950740021301 115, 122

D' Souza RN, Cavender A, Dickinson D, Roberts A, Letterio J. TGF-β1 is essential of homeostasis of the dentin pulp complex. Eur J Oral Sci, 1998: 106:185–191. 124

Erickson, H.(1993) : Tenascin-C, tenascin-R and tenascin-X : a family of talented proteins in search of functions. Curr. Opin. Cell Biol. 5:869 DOI: 10.1016/0955-0674(93)90037-Q 116

Felippe WT, Felippe MC, Rocha MJ. The effect of mineral trioxide aggregate on the apexification and periapical healing of teeth with incomplete root formation. Int Endod J (2006);39:2–9. 137

Finkelman RD, Mohan S, Jennings JC, Taylor AK, Jepsen S, Baylink DJ. Quantitation of growth factors IGF-I, SGF/IGF-II, and TGF-beta in human dentin. (1990) *J Bone Miner Res*. 1990: 5: 717–723. DOI: 10.1002/jbmr.5650050708 115, 117

Fuks AB, Michaeli Y, Sofer-Saks B, Shoshan S. Enriched collagen solution as a pulp dressing in pulpotomized teeth in monkeys. (1984) *Pediatr Dent*. 1984: 6: 243–247. 128

Furusawa M, Nakagawa K, Asai Y. Clinico-pathological studies on the tissue reactions of human pulp treated with various kinds of calcium phosphate ceramics. (1991) *Bull Tokyo Dent Coll*. 1991: 32: 111–120. 128

George, A., Gui, J., Jenkins, N.A., Gilbert, D.J., Copeland, N.G. and Veis, A. (1994): In situlocalization and chromosomal mapping of the AG1(Dmpl)gene. J. Histochem, Cytochem, 42:1527 114

George , A., Sabsay, B., Simonian, P.A.L. and Veis, .(1993):Characterization of a novel dentin matrix acidic phosphoprotein. J. Biol. Chem. 268:12624. 115

Goldberg, M. and Takagi, M. (1993):Dentine proteoglycans: composition, ultrastucture and functions. Histochem. J. 25:781. DOI: 10.1007/BF02388111

Goldberg M. and Lasfargues J.J. (1995): Pulp-dentinal complex revisited. J. Dentistry 23:15. DOI: 10.1016/0300-5712(95)90655-2 114

Goldberg M. , Septier D., Lecolle S., Vermelin L., Bissila-Mapahou P., Carreau J.P., Gritli A. and Bloch-Zupan A. (1995): Lipids in predentine and dentine. Connect. Tiss. Res. 32:427. 115

Goldberg M, Lacerda-Pinheiro S, Jegat N, Six N, Septier D, Priam F, Bonnefoix M, Tompkins K, Chardin H, Denbesten P, Veis A, Poliard A. The impact of bioactive molecules to stimulate tooth repair and regeneration as part of restorative dentistry. (2006) *Dent Clin North Am*. 2006: 50: 277–298. DOI: 10.1016/j.cden.2005.11.008 128

Goldberg M, Smith AJ. Cells and Extracellular Matrices of Dentin and Pulp: a Biological Basis for Repair and Tissue Engineering. Crit Rev Oral Biol Med (2004);15:13–27. 137

Gorter de Vries, I., Quartier, E., Van Steirteghem. A., Boute , P., Coomans. D. and Wisse, E. (1986): Characterization and immunological localization of dentine phosphoprotein in rat and bovine teeth. Arch. Oral Biol. 31:57. DOI: 10.1016/0003-9969(86)90114-7 115

Gronthos S, Brahim J, Li W, Fisher LW, Cherman N, Boyde A, et al. Stem cell properties of human dental pulp stem cells. J Dent Res (2002);81:531–5. DOI: 10.1177/154405910208100806

Gronthos S, Mankani M, Brahim J, Robey PG, Shi S. Postnatal human dental pulp stem cells (DPSCs) in vitro and in vivo. Proc Natl Acad Sci U S A 2000;97:13625–30.

Gurdon, J.B. (1992) : The generation of diversity and pattern in animal development. Cell 68:185. DOI: 10.1016/0092-8674(92)90465-O 122

Hanks CT, Sun ZL, Fang DN, Edwards CA, Wataha JC, Ritchie HH, Butler WT. Cloned 3T6 cell line from CD-1 mouse fetal molar dental papillae. (1998) *Connect Tissue Res*. 1998: 37: 233–249. DOI: 10.3109/03008209809002442 124

Hao et al. (2002), Hao J, Narayanan K, Ramachandran A, He G, Almushayt A, Evans C, George A. Odontoblast cells immortalized by telomerase produce mineralized dentin-like tissue both in vitro and in vivo. *J Biol Chem*. 2002: 277: 19976–19981 DOI: 10.1074/jbc.M112223200 123

Harbert H. One-step apexification without calcium hydroxide. J Endod (1996);22:690–2. DOI: 10.1016/S0099-2399(96)80066-5 137

Hachmeister DR, Schindler WG, Walker WA, 3rd, Thomas DD. The sealing ability and retention characteristics of mineral trioxide aggregate in a model of apexification. J Endod 2002;28:386–90. DOI: 10.1097/00004770-200205000-00010 137

Heikinheimo K, Happonen RP, Miettinen PJ, Ritvos O. Transforming growth factor beta 2 in epithelial differentiation of developing teeth and odontogenic tumors. (1027) *J Clin Invest*. 1993: 91:1019–1027. DOI: 10.1172/JCI116258 117

Heikinheimo K. Stage-specific expression of decapentaplegic-Vg-related genes 2, 4, and 6 (bone morphogenetic proteins 2, 4, and 6) during human tooth morphogenesis. (1994) *J Dent Res*. 1994: 73: 590–597. 117

Heritier M, Dangleterre M, Bailliez Y. Ultrastructure of a new generation of odontoblasts in grafted coronal tissues of mouse molar tooth germs. Arch Oral Biol (1989);34:875–83 DOI: 10.1016/0003-9969(89)90144-1 133

Heyeraas K.J. and Kvinnsland I. (1993):Tissue pressure and blood flow in pulpal inflammation. Proc. Finn. Dental. Soc. 88:393. 119

Heyeraas, K.J. (1990): Interstitial fluid pressure and transmicrovascular fluid flow. In Dynamic Aspects of Dental Pulp (Eds, R. Inoki, T. Kudo and L.M. Olgart). Charman and Hall, London, pp 189:198. 119

Heyeraas Tonder KJ, Kvinnsland I. Micropuncture measurements of interstitial fluid pressure in normal and inflamed dental pulp in cats. (1983) *J Endodont*. 1983: 9:105–109 DOI: 10.1016/S0099-2399(83)80106-X 125

Heys, D.R., Cox, C.F., Heys, R.J. and Avery, J.K. (1981):Histological considerations of direct pulp capping agents. J. Dent. Res. 60:1371. 120

Holtgrave EA, Donath K.Response of odontoblast-like cells to hydroxyapatite ceramic granules. (1995) *Biomaterials*. 1995: 16: 155 159. DOI: 10.1016/0142-9612(95)98280-R 128

Holz J, Baume IJ (1973):Esais biologiquer relaifs a la compatibilite des products d'obruration iner-mediaire avec l'organe pulpodentinaire, Revue mens Suisse Odonto-Stomatologie 83:1362. 114, 117

Horsted P, Sandergaard B, Thylstrup A, El Attar K, Fejerskov O. A retrospective study of di-rect pulp capping with calcium hydroxide compounds. Endod Dent Traumatol (1985);1:29–34. DOI: 10.1111/j.1600-9657.1985.tb00555.x 126

Huang GT, Sonoyama W, Liu Y, Liu H, Wang S, Shi S. The hidden treasure in apical papilla: the potential role in pulp/dentin regeneration and bioroot engineering. J Endod (2008);34:645–51. DOI: 10.1016/j.joen.2008.03.001 137

Imai M, Hayashi Y. Ultrastructure of wound healing following direct pulp capping with calcium-beta-glycerophosphate (Ca-BGP). (1993) *J Oral Pathol Med*. 1993: 22: 411–417. 128

Inoue T, Shimono M. Repair dentinogenesis following transplantation into normal and germ-free animals. Proc Finn Dent Soc (1992);88 Suppl 1:183–94. 132

Iohara K, Nakashima M, Ito M, Ishikawa M, Nakasima A, Akamine A. Dentin regeneration by dental pulp stem cell therapy with recombinant human bone morphogenetic protein 2. J Dent Res 2004;83:590–5. DOI: 10.1177/154405910408300802 124, 129

Ishizeki K, Fujiwara N, Nawa T. Morphogenesis of mineralized tissues induced by neona-tal mouse molar pulp isografts in the spleen. (1989) *Arch Oral Biol*. 1989: 34: 465–473. DOI: 10.1016/0003-9969(89)90126-X 132

Jacobsen, I., and Zachrisson. B.U (1975): Repair characteristics of root fra ctures in permanent anterior teeth. Scand. J. Dent. Res., 83:355. DOI: 10.1111/j.1600-0722.1975.tb00449.x 120

Jean AH, Pouezat JA, Daculsi G. Pulpal response to calcium phosphate materials, In vivo study of calcium phosphate materials in endodontics. (1993) *Cells Mater*. 1993: 3: 193–199. 128

Jepsen , S., Schiltz P., Strong D.D., Scharla S.H., Snead M.L. and Fiukelman R.D. (1992) : Trans-furming growth factor-â, mRNA in neonatal ovine molars visualired by in situ hybridization : potential role of the stratum intermedium. Archs Oral Biol. 37:645. DOI: 10.1016/0003-9969(92)90127-T 117

Jepsen S, Albers HK, Fleiner B, Tucker M, Rueger D. Recombinant human osteogenic protein-1 induces dentin formation: an experimental study in miniature swine. (1997) *J Endod.* 1997: 23: 378–382. DOI: 10.1016/S0099-2399(97)80187-2 136

Jernvall, J., Kettunen, P., Karavanova, I., Martin, L.B. and Thesleff, I. (1994) : Evidence for the role of the enamel knot as a control center in mammalian tooth cusp formation : non-dividing cells express growth stimulating Fgf-4 gene. Int. J. Dev. Biol. 38:463. 116

Joseph B.K., Savage N.W., Young N.G., Cupta G.S., Breier B.H. and Waters M.J. (1993) Expression and regulation of Insulin-like growth factor-1 in the rat incisor. *Growth factors* **8**, 267–275. DOI: 10.3109/08977199308991572 118

Joseph B.K., Savage N.W., Young W.G. and Waters M.J.(1994). Prenatal expression of growth hormone receptor binding protein and insulin-like growth factor-I (IGF-I) in the enamel organ. Role for growth hormone and IGF-I in cellular differentiation during early tooh formation. Anat. Embryol. 189:489. DOI: 10.1007/BF00186823 118

Kakehashi S, Stanley HR, Fitzgerald RJ. The effects of surgical exposures of dental pulps in germ-free and conventional laboratory rats. (1965) *Oral Surg Oral Med Oral Pathol.* 1965: 20: 340–349. DOI: 10.1016/0030-4220(65)90166-0

Kalyva M, Papadimitriou S, Tziafas D. Transdentinal stimulation of tertiary dentine formation and intratubular mineralization by growth factors. *Intern. Endod J.* 2010:43:382–392. 126, 135

Kamata N, Fujimoto R, Tomonari M, Taki M, Nagayama M, Yasumoto S. Immortalization of human dental papilla, dental pulp, periodontal ligament cells and gingival fibroblasts by telomerase reverse transcriptase. (2004) *J Oral Pathol Med.* 2004: 33: 417–423. DOI: 10.1111/j.1600-0714.2004.00228.x 123

Kasugai S, Shibata S, Suzuki S, Susami T, Ogura H. Characterization of a system of mineralized-tissue formation by rat dental pulp cells in culture. (1993) *Arch Oral Biol.* 1993: 38: 769–777. DOI: 10.1016/0003-9969(93)90073-U 123

Katz R.W., Reddi, A.H., (1988) : Dissociative extraction and partial purification of osteogenin, bone inductive protein, from rat tooth matrix by heparin affinity chromatograph. Biochem Biophys Res Commun 157:1253. DOI: 10.1016/S0006-291X(88)81009-X 118

Kawai, T., Urist,MR., (1989) : Bovine tooth-derived bone morphogenetic protein. J. Dent. Res. 68:1069. 118

Kawase T, Orikasa M, Ogata S, Burns DM. Protein tyrosine phosphorylation induced by epidermal growth factor and insulin-like growth factor-I in a rat clonal dental pulp-cell line. (1995) *Arch Oral Biol.* 1995: 40: 921–929. DOI: 10.1016/0003-9969(95)00061-S 124

Kettunen P, Thesleff I. Expression and function of FGFs-4, -8 and -9 suggest functional redundancy and repetitive use as epithelial signals during tooth morphogenesis. Dev. Dyn. 1998:22:374–385. 116

Kim S. (1985): Regulation of pulpal blood flow. J. Dent. Res. 65:602. 120

Kim, S.(1990): Neurovascular interactions in the dental pulp in health and inflammation. JOE, 16:48. DOI: 10.1016/S0099-2399(06)81563-3 118, 119, 125

Kinoshita, Y. (1979): Incorporation of serum albumin into the developing dentine and enamel matrix in the rabbit incisor. Calcif. Tissue Int. 29:41. DOI: 10.1007/BF02408054 115

Kollar EJ(1983): Epithelial-mesenchymal interaction in the mammalian intergument:tooth development as a model for instructive induction. In: Sawer RM, Fallou JF, eds, Epithelial Mesenchymal Interactions in Development. New York, USA:Praeger, pp 27–56. 114

Koling A, Rask-Andersen H. Membrane structures in the pulp-dentin border zone. A freeze-fracture study of demineralized human teeth. (1984) *Acta Odontol Scand*. 1984: 42: 73–84. 119

Kronmiller, J.E., Upholt, W.B., and Kollar, E.J. (1991)a : Expression of epidermal growth factor mRNA in the developing mouse mandibular process. Arch. Oral Biol. 36:405. DOI: 10.1016/0003-9969(91)90129-I 116

Kubler, D, Lesot, H, Ruch, J.V. (1988) : Temporo-spatial distribution of matrix and microfilament components during odontoblast and ameloblast differentiation. Roux Archives of Developmental Biology 197:212.. DOI: 10.1007/BF02439428 116

Kuo MY, Lan WH, Lin SK, Tsai KS, Hahn LJ. Collagen gene expression in human dental pulp cell cultures. (1992) *Arch Oral Biol*. 1992: 37: 945–952. DOI: 10.1016/0003-9969(92)90066-H 121

Langeland, K.(1959): Histologic evaluation of pulp reactions to operative procedures. Oral Surg. 12:1235. 120

Langeland, K.; Dowden. W.E.; Tronstad, L.; and Langeland, K. (1971): Human pulp changes of iatrogenic origin. Oral Surg. 32:943. DOI: 10.1016/0030-4220(71)90183-6 120

Larmas, M.,(1986): Response of pulpodentinal complex to caries attach. Proc. Finn. Soc. 82:298. 120

Langeland K. Tissue response to dental caries. (1987) *Endod Dent Traumatol*. 1987: 3: 149–171 DOI: 10.1111/j.1600-9657.1987.tb00619.x 120

Lanjia Y, Yuhao G, White FH. Bovine-bone morphogenetic protein-induced dentinogenesis. Clin. Orthop. Relat. Res. (1993):295:305–312. 129

Larmas M. Response of pulpodentinal complex to caries attack. (1986) *Proc Finn Dent Soc*. 1986: 82: 298–304.

Lesot H, Osman M, Ruch JV. Immunofluorescent localization of collagens, fibronectin, and laminin during terminal differentiation of odontoblasts. (1981) *Dev Biol*. 1981: 82: 371–381. DOI: 10.1016/0012-1606(81)90460-7 115, 116, 117, 123

Lesot H, Begue-Kirn C, Kubler MD, Meyer JM, Smith AJ, Cassidy N, Ruch JV. Experimental induction of odontoblast differentiation and stimulation during reparative processes. (1993) *Cells Mater* 1993: 3: 201–217 119, 123

Lesot H, Smith AJ, Tziafas D, Begue-Kirn C, Cassidy N, Ruch JV. Biologically active molecules and dental tissue repair: a comparative review of reactionary and reparative dentinogenesis with the induction of odontoblast differentiation in vitro. (1994) *Cells Mater*. 1994: 4: 199–218 119, 123, 129

Liang RF, Nishimura S, Maruyama S, Hanazawa S, Kitano S, Sato S. Effects of transforming growth factor-beta and epidermal growth factor on clonal rat pulp cells. (1990) *Arch Oral Biol*. 1990: 35: 7–11. DOI: 10.1016/0003-9969(90)90106-K 124

Linde, A., (1989): Dentin matrix proteins:composition and possible functions in calcification . Anat. Rec. 224:154. 114

Linde, A., and Goldberg, M. (1993): Dentinogenesis. Crit. Rev. Oral Biol. Med, 4:679. 114, 115

Liu H, Li W, Gao C. Dentonin, a fragment of MEPE, dental pulp stem cell proliferation. (2004) *J Dent Res*, 2004:83:496–9. 128

Lyons KM, Pelton RW, Hogan BLM. Organogenesis and pattern formation in the mouse: RNA distribution patterns suggest a role for bone morphogenetic protein-2A (BMP-2A). (1990) *Development* 1990: 109: 833–844 117

Luostarinen V, Ronning O. Differences in the osteoinductive potential of transplanted isogeneic dental structures of the rat. (1977) *Acta Anat (Basel)*. 1977: 99: 76–83 DOI: 10.1159/000144837 132

Magloire H, Joffre A, Grimaud JA, Herbage D, Couble ML, Chavrier C, et al. Synthesis of type I collagen by human odontoblast-like cells in explant culture: light and electron microscope immunotyping. Cell Mol Biol Incl Cyto Enzymol (1981);27:429–35 120, 121

Magloire H, Joffre A, Hartmann DJ. Localization and synthesis of type III collagen and fibronectin in human reparative dentine. Immunoperoxidase and immunogold staining. (1988) *Histochemistry*. 1988: 88: 141–149. DOI: 10.1007/BF00493296 122, 123, 125

Magloire H, Joffre A, Bleicher F. An in vitro model of human dental pulp repair. J Dent Res (1996);75:1971–8. DOI: 10.1177/00220345960750120901 124

Magloire H, Romeas A, Melin M, Couble ML, Bleicher F, Farges JC. Molecular regulation of odontoblast activity under dentin injury. (2001) *Adv Dent Res*. 2001: 15: 46–50. 124

Mac Dougall M. Dentine phosphoprotein in dentin development: implications in dentinogenesis imperfecta. (1992) *Proc Finn Dent Soc*. 1992: 88: 195- 204. 121

MacDougall M, Thiemann F, Ta H, Hsu P, Chen LS, Snead ML. Temperature sensitive simian virus 40 large T antigen immortalization of murine odontoblast cell cultures: establishment of clonal odontoblast cell line. Connect Tissue Res 1995;33:97–103. DOI: 10.3109/03008209509016988 121, 123

McDougall M, Simmons D, Luan X, Nyggeder J, Feng J, Gu TT. Dentin phosphoprotein and dentin sialoprotein are cleaved products expressed from a single transcript coded by a gene on human chromosome 4. (1997) *J Biol Chem*. 1997: 272: 835–842. DOI: 10.1074/jbc.272.2.835 124

MacKenzie, A., Leeming, G., Jowett, A.K., Ferguson, M.W and Sharpe, P.T. (1991) : The homeobox gene 7.1 has specific regional and temporal expression patterns during early murine craniofacial embryogenesis, especially tooth development in vivo and in vitro. Development 111:269. 116

MacKenzie, A., Ferguson, M., WJ, sharpe PT (1992) : Expression patterns of the homeobox gene, Hox-8 in mouse embryo suggest a role in specifying tooth initiation and shape. Development 115:403. 116

Meister, E., Jr., Lommel, T.J., and Gerstein, H.(1980):Diagnosis and possible causes of vertical root fractures. Oral Surg. 49:243. DOI: 10.1016/0030-4220(80)90056-0 120

Meyer JM, Lesot H, Staubli A, Ruch JV. Immunoperoxidase localization of fibronectin during odontoblast differentiation. An ultrastructural study. (1989) *Biol Struct Morphog*. 1989: 2: 19–24. 117

Mitsiadis T.A., Dicou E., Joffre A. and Magoire H. (1992). Immunohistochemical localization of nerve growth factor (NGF) and NGF-receptor (NGF-r) in the developing first molar tooth of the rat. Differentiation 49:47. DOI: 10.1111/j.1432-0436.1992.tb00768.x 118

Mitsiadis T., Couble P., Dicou E., Joffre A. and Magloire H. (1993) Patterns of nerve growh factor (NGF), proNGF and p75NGF receptor expression in the rat incisor: comparison with expression in the molar. *Differentiation* **54,** 161–175. 118

Mitsiadis T.A. and Luukko K. (1995). Neurotrophins in odontogenesis. Int. J. Dev. Biol. 39:195. 118

Miura M, Gronthos S, Zhao M, Lu B, Fisher LW, Robey PG, et al. SHED: stem cells from human exfoliated deciduous teeth. Proc Natl Acad Sci U S A (2003);100:5807–12. DOI: 10.1073/pnas.0937635100

Mjor, I.A. (1977): Histologic demonstration of bacterial subjacent to dental restorations. Scand. J. Dent. Res. 85:169. DOI: 10.1111/j.1600-0722.1977.tb00550.x 120, 121

Mjor, I. A. (Ed) (1983): Dentin and pulp. In Reaction Patterns in Human Teeth. CRC Press, Florida, pp.63:156. 134

Moullec, N. (1978): Contribution a l'etude morphologigue du development normal dew dents chez la souris causalite de la differenciation dentaire. Archives d'Anatomy. Histologie et Embrologie, Normales et Experimentales 61:161. 114

Murray PE, Matthews JB, Sloan AJ, Smith AJ. Analysis of incisor pulp cell populations in Wistar rats of different ages. (2002) *Arch Oral Biol.* 2002: 47: 709–715. DOI: 10.1016/S0003-9969(02)00055-9 125

Murray PE, Stanley HR, Matthews JB, Sloan AJ, Smith AJ. Ageing human odontometric analysis. (2002) *Oral Surg Oral Med Oral Pathol Oral Radiol* 2002: 93: 474–482 DOI: 10.1067/moe.2002.120974

Murray PE, Smith AJ. Saving pulps-a biological basis. An overview. (2002) *Prim Dent Care.* 2002: 9: 21–26. 125

Nair PN, Duncan HF, Pitt Ford TR, Luder HU. Histological, ultrastructural and quantitative investigations on the response of healthy human pulps to experimental capping with mineral trioxide aggregate: a randomized controlled trial. Int Endod J (2008);41:128–50. DOI: 10.1111/j.1365-2591.2007.01329.x 126

Nakashima M. Dentin induction by implants of autolyzed antigen-extracted allogeneic dentin on amputated pulps of dogs. (1989) *Endod Dent Traumatol.* 1989: 5: 279–286. DOI: 10.1111/j.1600-9657.1989.tb00374.x 128

Nakashima M. Establishment of primary cultures of pulp cells from bovine permanent incisors. Arch Oral Biol (1991);36:655–63. DOI: 10.1016/0003-9969(91)90018-P 121

Nakashima M. The effects of growth factors on DNA synthesis, proteoglycan synthesis and alkaline phosphatase activity in bovine dental pulp cells. Arch Oral Biol (1992);37:231–6. DOI: 10.1016/0003-9969(92)90093-N 123

Nakashima M, Nagasawa H, Yamada Y, Reddi H. Regulatory role of transforming growth factor-β bonemorphogenetic protein-2 and protein -4 on gene expression of extracellular matrix proteins and differentiation on dental pulp cells. (1994) *Dev Biol*1994: 162: 18–28. DOI: 10.1006/dbio.1994.1063 129

Nakashima M. Induction of dentin formation on canine amputated pulp by recombinant human bone morphogenetic proteins (BMP)-2 and -4. (1994) *J Dent Res*. 1994: 73: 1515–1522. 129

Nakashima M. Induction of dentine in amputated pulp of dogs by recombinant human bone morphogenetic proteins-2 and -4 with collagen matrix. (1994) *Arch Oral Biol*. 1994: 39: 1085–1089.

Nakashima M, Reddi AH. The application of bone morphogenetic proteins to dental tissue engineering. (2003) *Nat Biotechnol*. 2003: 21: 1025–1032. 129

Nakashima M, Tachibana K, Iohara K, Ito M, Ishikawa M, Akamine A. Induction of reparative dentin formation by ultrasound-mediated gene delivery of growth/differentiation factor 11. (2003) *Hum Gene Ther*. 2003: 14: 591–597. DOI: 10.1089/104303403764539369 129

Nakashima M. Bone morphogenetic proteins in dentin regeneration for potential use in endodontic therapy. (2005) *Cytokine Growth Factor Rev*. 2005: 16: 369–376. DOI: 10.1016/j.cytogfr.2005.02.011

Nakashima M, Iohara K, Zheng L. Gene therapy for dentin regeneration with bone morphogenetic proteins. (2006) *Curr Gene Ther*. 2006: 6: 551–560.

Niswander, L. and Martin, G.R. (1992) : Fgf-4 expression during gastrulation, myogenesis, limb and tooth development in the mouse. Development 114:755. 116

Nor, J.E. (2006) Tooth regeneration in operative dentistry. Oper Dent., 2006: 31-6: 633-42. 136

Oguntebi BR, Dover MS, Franklin CJ, Tuwaijri AS. The effect of collagen and indomethacin on inflamed dental pulp wounds of baboon teeth. (1988) *Oral Surg Oral Med Oral Pathol*. 1988: 65: 233–239. DOI: 10.1016/0030-4220(88)90172-7 128

Olgart LM. Functions of peptidergic nerves. In:Inoki, R., Kudo, T., Olgart, L.M.(eds). (1990) *Dynamic aspects of dental pulp*. Chapman and Hall,London, 1990:349. 119

Olgart, L.M. (1992): Involvement of sensory nerves in demodynamic reactions. Proc. Finn. Dent. Soc. 88:403. 119

Oosterwegel, M., Van De Wetering, M., Timmerman, J., Kruisbeed, A., Destree, O., Meijlink, F and Clevers, H. (1993) : Differential expression of the HMG box factors TCF-1 and LEF-1 during murine embryogenesis. Development 118:439. 116

Ohshima H. Ultrastructural changes in odontoblasts and pulp capillaries following cavity preparation in rat molars. (1990) *Arch Histol Cytol*. 1990: 53: 423–438. 120

. Osman A, Ruch JV. Repartition topographique des mitoses dans l' incisive et la 1ere molaire inferieure de l' embryon de souris, J. Biol. Buccale 1967: 8: 4:331. 117

Papaccio G, Graziano A, d'Aquino R, Graziano MF, Pirozzi G, Menditti D, et al. Long-term cryopreservation of dental pulp stem cells (SBP-DPSCs) and their differentiated osteoblasts: a cell source for tissue repair. J Cell Physiol 2006;208:319–25. DOI: 10.1002/jcp.20667

Parish CA, Smrcka AV, Rando RR. Functional significance of beta gamma-subunit carboxymethylation for the activation of phospholipase C and phosphoinositide 3-kinase. (7727) *Biochemistry*. 1995: 34: 7722–7727. DOI: 10.1021/bi00023a019 124

Partanen, A.M., and Thesleff, I. (1987) : Localization and quantitation of 125l-epidermal growth factor brinding in mouse embryonic tooth and other embryonic tissues at different developmental stages. Dev.Biol. 120:186. DOI: 10.1016/0012-1606(87)90117-5 116

Pashley DH. Dentin-predentin complex and its permeability: physiologic overview. (1985) *J Dent Res*. 1985: 64: 613–620. 125

Prescott RS, Alsanea R, Fayad MI, Johnson BR, Wenckus CS, Hao J, et al. In vivo generation of dental pulp-like tissue by using dental pulp stem cells, a collagen scaffold, and dentin matrix protein 1 after subcutaneous transplantation in mice. J Endod 2008;34:421–6. DOI: 10.1016/j.joen.2008.02.005 137

Qin C, Brunn JC, Cadena E, Ridall A, Tsujigiwa H, Nagatsuka H, Nagai N, Butler WT. The expression of dentin sialophosphoprotein gene in bone. (2002) *J Dent Res*. 2002: 81:392–394. DOI: 10.1177/154405910208100607 124

Ritchie H.H., Pinero G.J., Hou H. and Butler W.T. (1995): Molecular analysis of rat dentin sialoprotein. Connect. Tissu. Res. 33:395. DOI: 10.3109/03008209509016985 115

Robey, B.G.(1989): The biochemisry of bone . Endocrinol Metab. Clinics of North Am. 18:859. 114

Ruch JV. Odontoblast differentiation and the formation of the odontoblast layer. (1985) *J Dent Res*. 1985: 64: 489–498. 114, 116

Ruch JV, Lesot H, Begue-Kirn C. Odontoblast differentiation. (1995) *Int J Dev Biol*. 1995: 39: 51–68. 114, 116

Rutherford RB, Wahle J, Tucker M, Rueger D, Charette M. Induction of reparative dentine formation in monkeys by recombinant human osteogenic protein-1. (1993) *Arch Oral Biol*. 1993: 38: 571–576. DOI: 10.1016/0003-9969(93)90121-2 129, 136

Rutherford B, Spangberg L, Tucker M, Charette M. Transdentinal stimulation of reparative dentine formation by osteogenic protein-1 in monkeys. (1995) *Arch Oral Biol*. 1995: 40: 681–683. DOI: 10.1016/0003-9969(95)00020-P 125, 126, 135

Rutherford BR: Regeneration of the Pulp-Dentin complex: In Lynch SE, Genco RJ, Marx RE eds: (1999) *Tissue engineering. Applications in maxillofacial surgery and periodontics*, Quintessence Publishing Co, Inc, Chicago, 1999: 185–199. 130

Rutherford RB. BMP-7 gene transfer to inflamed ferret dental pulps. (2001) *Eur J Oral Sci.* 2001: 109: 422–424. DOI: 10.1034/j.1600-0722.2001.00150.x 129

Sasaki T, Kawamata-Kido H. Providing an environment for reparative dentine induction in amputated rat molar pulp by high molecular-weight hyaluronic acid. (1995) *Arch Oral Biol.* 1995: 40: 209–219. DOI: 10.1016/0003-9969(95)98810-L 128

Satokata I, Maas R. Msx1 deficient mice exhibit cleft palate and abnormalities of craniofacial and tooth development. Nat Genet. 1994 Apr; 6(4): 348–56. 116

Schroder U. Effects of calcium hydroxide-containing pulp-capping agents on pulp cell migration, proliferation, and differentiation. (1985) *J Dent Res.* 1985: 64 (Spec No): 541–548.

Searls, J.C. (1975): Radioautographic evaluation of changes induced in the ra incisor b high-speed caviy preparatiion. J. Dent. Res., 54:174. 120

Senzaki H. A histological study of reparative dentinogenesis in the rat incisor after colchicine administration. (1980) *Arch Oral Biol.* 1980: 25: 737–743. DOI: 10.1016/0003-9969(80)90127-2 125

Seux D, Couble ML, Hartmann DJ, Gauthier JP, Magloire H. Odontoblast-like cytodifferentiation of human dental pulp cells in vitro in the presence of a calcium hydroxide-containing cement. (1991) *Arch Oral Biol.* 1991: 36: 117–128. DOI: 10.1016/0003-9969(91)90074-5 123

Shi S, Gronthos S. Perivascular niche of postnatal mesenchymal stem cells in human bone marrow and dental pulp. J Bone Miner Res (2003);18:696–704. DOI: 10.1359/jbmr.2003.18.4.696

Shiba H, Nakamura S, Shirakawa M, Nakanishi K, Okamoto H, Satakeda H, Noshiro M, Kamihagi K, Katayama M, Kato Y. Effects of basic fibroblast growth factor on proliferation, the expression of osteonectin (SPARC) and alkaline phosphatase, and calcification in cultures of human pulp cells. (1995) *Dev Biol.* 1995: 170: 457–466. DOI: 10.1006/dbio.1995.1229 123

Shirakawa M, Shiba H, Nakanishi K, Ogawa T, Okamoto H, Nakashima K, Noshiro M, Kato Y. Transforming growth factor-beta-1 reduces alkaline phosphatase mRNA and activity and stimulates cell proliferation in cultures of human pulp cells. (1514) *J Dent Res.* 1994: 73: 1509–1514. DOI: 10.1177/00220345940730090501 123

Silverstone, L.M. and Mjor , I.A (1988): Dental caries: caries of dentin In:Horsted-Bindsler P. and Mjor , I. A. (eds) Modern concepts in operative dentistry, p. 45:58. 120

Six N, Lasfargues JJ, Goldberg M. In vivo study of the pulp reaction to Fuji IX, a glass ionomer cement. (2000) *J Dent*. 2000: 28: 413–22. 129

Slavkin HC, Zeichner-David M, Siddiqui M. Molecular aspects of tooth morphogenesis and differentiation. Mol Aspects of Med. 1981:4:73-86. 114, 116

Sloan AJ, Smith AJ. Stimulation of the dentine-pulp complex of rat incisor teeth by transforming growth factor-beta isoforms 1–3 in vitro. (1999) *Arch Oral Biol*. 1999: 44: 149–156. DOI: 10.1016/S0003-9969(98)00106-X 124

Sloan AJ, Smith AJ. Stem cells and the dental pulp:potential roles in dentine regeneration and repair. Oral Dis (2007);13:151–7. DOI: 10.1111/j.1601-0825.2006.01346.x 137

Smith AJ: Pulpal responses to caries and dental repair. (2002) *Caries Res* 2002;36:223–232. DOI: 10.1159/000063930 125, 126, 135

Smith AJ, Tobias RS, Plant CG, Browne RM, Lesot H, Ruch JV. In vivo morphogenetic activity of dentine matrix proteins. (1990) *J Biol Buccale*. 1990: 18: 123–129. 129

Smith AJ, Cassidy N, Perry H, Begue-Kirn C, Ruch JV, Lesot H. Reactionary dentinogenesis. (1995) *Int J Dev Biol*. 1995: 39: 273–280. 121, 134

Smith AJ, Murray PE, Sloan AJ, Matthews JB, Zhao S. Trans-dentinal stimulation of tertiary dentinogenesis. Adv Dent Res. (2001): 15: 51–54. DOI: 10.1177/08959374010150011301 115, 126, 135

Sonoyama W, Liu Y, Fang D, Yamaza T, Seo BM, Zhang C, et al. Mesenchymal stem cell-mediated functional tooth regeneration in swine. PLoS ONE (2006);1:e79. DOI: 10.1371/journal.pone.0000079 137

Sonoyama W, Liu Y, Yamaza T, Tuan RS, Wang S, Shi S, et al. Characterization of the apical papilla and its residing stem cells from human immature permanent teeth: a pilot study. J Endod (2008);34:166–71. DOI: 10.1016/j.joen.2007.11.021

Stanley H.R. (1981): Human Pulp Response to Restorative Dental Procedures. Storter Printing Co, Grainesville.

Stanley, H.R. (1993): Effects of dental restorative materials. JADA, 124:76. 120

Stanley HR. Pulp capping: conserving the dental pulp–can it be done? Is it worth it? Oral Surg Oral Med Oral Pathol (1989);68:628–39. DOI: 10.1016/0030-4220(89)90252-1 128

Stanley HR, Lundy T. Dycal therapy for pulp exposures. (1972) *Oral Surg Oral Med Oral Pathol*. 1972: 34: 818–827. DOI: 10.1016/0030-4220(72)90300-3 114, 125

Suwa F, Yang L, Ohta Y, Fang YR, Ike H, Deguchi T. Ability of hydroxyapatite-bone morphogenetic (corrected from morphologenetic) protein (BMP) complex to induce dentin formation in dogs. (1993) *Okajimas Folia Anat Jpn*. 1993: 70: 195–201. 129

Sveen OB, Hawes RR. Differentiation of new odontoblasts and dentine bridge formation in rat molar teeth after tooth grinding. (1968) *Arch Oral Biol*. 1968: 13: 1399–409. DOI: 10.1016/0003-9969(68)90022-8 125, 134

Takeda T, Tezuka Y, Horiuchi M, Hosono K, Iida K, Hatakeyama D, et al. Characterization of dental pulp stem cells of human tooth germs. J Dent Res (2008);87:676–81. DOI: 10.1177/154405910808700716

Takagi, M., Hishikawa, H., Hosokawa, Y. , Kagami, A. and Rahemtulla, F. (1990): Immunohistochemical localization of glycosamionoglycans and proteoglycans in predentin an dentin of rat incisors. H. Histochem. Cytochem 38:319. 115

Takei K, Inoue T, Shimono M, Yamamura T. An experimental study of dentinogenesis in autografted dental pulp in rats. (1988) *Bull Tokyo Dent Coll*. 1988: 29: 9–19.

Takei K, Inoue T, Shimono M, Yamamura T. Dentinogenic activity of autografted dental pulp tissue of rat incisor. (1987) *J Dent Res*. 1987: 66: 287–94. 132, 133

Ten Cate J. Reaction parer: session I. Odontoblasts. (1985) *J Dental Res*. 1985: 64: 549. 120

Thesleff I, Barrach HJ, Foidart JM, Vaheri A, Pratt RM, Martin GR. Changes in the distribution of type IV collagen, laminin, proteoglycan, and fibronectin during mouse tooth development. (1981) *Dev Biol*. 1981: 81: 182–192. DOI: 10.1016/0012-1606(81)90361-4 116

Thesleff I, Mackie E, Vainio S, Chiquet-Ehrismann R. Changes in the distribution of tenascin during tooth development. Development (1987);101:289–96. 116, 117

Thesleff, I.,Vaaitokari A (1992):The role of growth factors in determination and differentiation of the odontoblastic cell lineage Om:Kim S. Ed proceedings of Finnish Dental Society 20:357. 114

Thesleff I, Vaahtokari A, Kettunen P, Aberg T. Epithelial-mesenchymal signaling during tooth development. Conn Tissue Res. 1995:32:9-15. 116, 117

Thesleff I, Hurmerinta K. Tissue interactions in tooth development. Differentiation (1981);18:75–88. 114, 117

Thesleff I, Vaahtokari A. The role of growth factors in determination and differentiation of the odontoblastic cell lineage. (1992) *Proc Finn Dent Soc*. 1992: 88: 357–368. 117

Thibodeau B, Teixeira F, Yamauchi M, Caplan DJ, Trope M. Pulp revascularization of immature dog teeth with apical periodontitis. J Endod 2007; 33: 680–9. 137

Thomas HF, Kollar EJ. Differentiation of odontoblasts in grafted recombinants of murine epithelial root sheath and dental mesenchyme. Arch Oral Biol (1989);34:27–35 DOI: 10.1016/0003-9969(89)90043-5 133

Thonemann B, Schmalz G. Bovine dental papilla-derived cells immortalized with HPV 18 E6/E7. (2000) *Eur J Oral Sci.* 2000: 108: 432–441. DOI: 10.1034/j.1600-0722.2000.108005432.x 123

Tominaga H, Sasaki T, Higashi S. Ultrastuctural changes in odontoblasts during early development. (1984) *Bull Tokyo Dent Coll.* 1984: 25: 9–26. 114, 117

Torneck CD, Smith JS, Grindall P. Biologic effects of endodontic procedures on developing incisor teeth. 3. Effect of debridement and disinfection procedures in the treatment of experimentally induced pulp and periapical disease. Oral Surg Oral Med Oral Pathol 1973a;35:532–40. DOI: 10.1016/0030-4220(73)90012-1 137

Torneck CD, Smith JS, Grindall P. Biologic effects of endodontic procedures on developing incisor teeth. II. Effect of pulp injury and oral contamination. Oral Surg Oral Med Oral Pathol (1973)b;35:378–88. DOI: 10.1016/0030-4220(73)90076-5 137

Trowbridge HO. Pathogenesis of pulpitis resulting from dental caries. (1981) *J Endod.* 1981: 7: 52–60. DOI: 10.1016/S0099-2399(81)80242-7 120, 125

Tsukamoto Y, Fukutani S, Shin-Ike T, Kubota T, Sato S, Suzuki Y, Mori M. Mineralized nodule formation by cultures of human dental pulp-derived fibroblasts. (1992) *Arch Oral Biol.* 1992: 37: 1045–1055. DOI: 10.1016/0003-9969(92)90037-9 121, 123

Turner DF, Marfurt CF, Sattelberg C. Demonstration of physiological barrier between pulpal odontoblasts and its perturbation following routine restorative procedures: a horseradish peroxidase tracing study in the rat. (1989) *J Dent Res.* 1989: 68: 1262–1268. 119

Tziafas D. Mechanisms controlling secondary initiation of dentinogenesis. (1994) *Int Endod J.* 1994: 27: 61–74. 119, 122

Tziafas D. Basic mechanisms of cytodifferentiation and dentinogenesis during dental pulp repair. Int J Dev Biol (1995);39:281–90. 121, 131

Tziafas D. Reparative Dentinogenesis. A monograph on the dentonogenic potential of dental pulp. University Studio Press, Thessaloniki, Greece. (1997). 130

Tziafas D. The future role of a molecular approach to pulp-dentinal regeneration. (2004) *Caries Res.* 2004: 38: 314–320 DOI: 10.1159/000077771 134

Tziafas D, Amar S, Staubli A, Meyer JM, Ruch JV. Effects of glycosaminoglycans on in vitro mouse dental cells. Arch Oral Biol (1988);33:735–40. DOI: 10.1016/0003-9969(88)90007-61 123

Tziafas D, Alvanou A, Kaidoglou A. Dentinogenic activity of allogenic plasma fibronectin on dog dental pulp. (1992) *J Dent Res*, 1992 :71:1189–1195. 130

Tziafas D, Smith AJ and Lesot H: Designing new treatment strategies in vital pulp therapy. (2000) *J Dentist* 2000: 28: 77–92. DOI: 10.1016/S0300-5712(99)00047-0 122, 129, 130, 134

Tziafas D, Kolokuris I. Inductive influences of demineralized dentin and bone matrix on pulp cells: an approach of secondary dentinogenesis. J Dent Res (1990);69:75–81. 130

Tziafas D, Kolokuris I, Alvanou A, Kaidoglou K. Short-term dentinogenic response of dog dental pulp tissue after its induction by demineralized or native dentine, or predentine. (1992) *Arch Oral Biol*. 1992: 37: 119–128.
DOI: 10.1016/0003-9969(92)90007-U 131

Tziafas D, Alvanou A, Panagiotakopoulos N, Smith AJ, Lesot H, Komnenou A, Ruch JV. Induction of odontoblast-like cell differentiation in dog dental pulps after in vivo implantation of dentine matrix components. (1995) *Arch Oral Biol*. 1995: 40: 883–893. DOI: 10.1016/0003-9969(95)00069-2 122, 130

Tziafas D, Alvanou A, Papadimitriou S, Gasic J, Komnenou A. Effects of recombinant basic fibroblast growth factor, insulin-like growth factor-II and transforming growth factor-beta 1 on dog dental pulp cells in vivo. (1998) *Arch Oral Biol*. 1998: 43: 431–444. 132

Tziafas D, Belibasakis G, Veis A, Papadimitriou S. Dentin regeneration in vital pulp therapy: design principles. (2001) *Adv Dent Res*. 2001: 15: 96–100. DOI: 10.1177/08959374010150012501 136

Tziafas D, Kalyva M. and Papadimitriou S. Experimental dentin-based approaches to tissue regeneration in vital pulp therapy. Con Tissue Res, (2002):43:391–395. DOI: 10.1080/713713458

Tziafas D, Kolokuris I. Inductive influences of demineralized dentin and bone matrix on pulp cells: an approach of secondary dentinogenesis. (1990) *J Dent Res*. 1990: 69: 75–81.

Tziafas D, Lambrianidis T, Beltes P. Inductive effect of native dentin on the dentinogenic potential of adult dog teeth. (1993) *J Endod*. 1993: 19: 116–122. DOI: 10.1016/S0099-2399(06)80505-4 130

Tziafas D, Panagiotakopoulos N, Komnenou A. Immunolocalization of fibronectin during the early response of dog dental pulp to demineralized dentine or calcium hydroxide-containing cement. Archs Oral Biol (1995):40:23–31. DOI: 10.1016/0003-9969(94)00148-5 132

Tziafas D, Papadimitriou S. Role of exogenous TGF-beta in induction of reparative dentinogenesis in vivo. *Eur J Oral Sci*. 1998: 106 (Suppl 1): 192–196. 132

Vaahtokari, A., Vainio, S., Thesleff, I. (1991) : Associations between transforming growth factor âl RNA expression and epithelial-mesenchymal interactions during tooth morphogenesis. Development. 113:985. 117

Van Genderen, C., Okamura, R.M., Farinas, I., QUO, R., Parslow, T.G., Bruhn, L. and Grosschedl, R. (1994) : Develmopment of several organs that require inductive epithelial-mesenchymal interactions is impaired in LEF-1 deficient mice. Genes Dev. 8:2691 DOI: 10.1101/gad.8.22.2691 116

Vainio, S., Jalkanen, M., and Thesleff, I., (1989) : Syndecan and tenascin expression is induced by epithelial-mesenchymal interactions in embryonic tooth mesencyme. J.Cell Biol. 108:1945. DOI: 10.1083/jcb.108.5.1945 116

Vainio S, Karavanova I, Jowett A, Thesleff I. Identification of BMP-4 as a signal mediating secondary induction between epithelial and mesenchymal tissues during early tooth development. (1993) *Cell*. 1993: 75: 45–58. 117

Van Hassel, H.J.(1971): Physiology of the human dental pulp Oral Surg. 32:126. DOI: 10.1016/0030-4220(71)90258-1 119

Veis A. The role of dental pulp-thoughts on the session on pulp repair processes. (1985) *J Dent Res*. 1985: 64: 552–554. 114

Veis, A. (1993): Mineral-matrix inteactions in bone and dentin. J. Bone Miner. Res. 8:5493. DOI: 10.1002/jbmr.5650081312 114

Veis, A. and Perry, A. (1967) : The phosphoprotein of the dentin matrix. Biochemisty 6:2409. DOI: 10.1021/bi00860a017 115

Veron MH, Couble ML, Caillot G, Hartmann DJ, Magloire H. Expression of fibronectin and type I collagen by human dental pulp cells and gingiva fibroblasts grown on fibronectin substrate. Arch Oral Biol (1990);35:565–9 DOI: 10.1016/0003-9969(90)90089-S 123

Yamamura T. Differentiation of pulpal cells and inductive influences of various matrices with reference to pulpal wound healing. (1985) *J Dent Res*. 1985: 64: 530–540. 119, 125, 132, 133

Young W.G. (1995). Growth hormone and insulin-like growth factor-I in odontogenesis. Int. J. Dev. Biol. 39:263 118

Young CS, Terada S, Vacanti JP, Honda M, Bartlett JD, Yelick PC. Tissue engineering of complex tooth structures on biodegradable polymer scaffolds. J Dent Res (2002);81:695–700. DOI: 10.1177/154405910208101008 137

Yoshida T, Itoh T, Saitoh T, Sekine I. Histopathological study of the use of freeze-dried allogenic dentin powder and True Bone Ceramic as apical barrier materials. J Endod (1998);24:581–6. DOI: 10.1016/S0099-2399(98)80114-3 137

Yoshida T, Nakagama K, Asai Y. Ultrastructural studies of initial calcification on exposed human pulp applied with two kinds of hydroxyapatite ceramics. (1992) *Bull Tokyo Dent Coll*. 1992: 33: 51–9. 128

Yoshiba K, Yoshiba N, Iwaku M. Effects of antibacterial capping agents on dental pulps of monkeys mechanically exposed to oral microflora. (1995) *J Endod*. 1995: 21: 16–20. DOI: 10.1016/S0099-2399(06)80551-0

Yoshimine Y, Maeda K. Histologic evaluation of tetracalcium phosphate-based cement as a direct pulp-capping agent. (1995) *Oral Surg Oral Med Oral Pathol Oral Radiol Endod*. 1995: 79: 351–358. DOI: 10.1016/S1079-2104(05)80229-X 128

Yoshiba N, Yoshiba K, Iwaku M, Nakamura H, Ozawa H. A confocal laser scanning microscopic study of the immunofluorescent localization of fibronectin in the odontoblast layer of human teeth. (1994) *Arch Oral Biol*. 1994: 39: 395–400. DOI: 10.1016/0003-9969(94)90169-4 117

Yoshiba K, Yoshiba N, Nakamura H, Iwaku M, Ozawa H. Immunolocalization of fibronectin during reparative dentinogenesis in human teeth after pulp capping with calcium hydroxide. (1996) *J Dent Res*. 1996: 75: 1590–1597. DOI: 10.1177/00220345960750081101 122

Yoshida, S., and Massler, M., (1984): Pulpal response to dental caries N.Y. J. Dent. 34:21. 120

Wise GE, Fan W. Immunolocalization of transforming growth factor beta in rat molars. (1991) *J Oral Pathol Med*. 1991: 20: 74–80. DOI: 10.1111/j.1600-0714.1991.tb00893.x 117

Zach, L., (1972): Pulp lability and repair: Effect of restorative procedures. Oral Surg. 33:111. DOI: 10.1016/0030-4220(72)90215-0 120

Zhang W, Walboomers XF, Van Kuppevelt TH, Daamen WF, Van Damme PA, Bian Z, et al. In vivo evaluation of human dental pulp stem cells differentiated towards multiple lineages. J Tissue Eng Regen Med (2008);2:117–25. DOI: 10.1002/term.71

Zhang W, Abukawa H, Troulis MJ, Kaban LB, Vacanti JP, Yelick PC. Tissue engineered hybrid tooth-bone constructs. Methods (2009);47:122–8. DOI: 10.1016/j.ymeth.2008.09.004 137

Zussman WV. Osteogenic activity of odontoblasts in transplanted tooth pulps. J Dent Rest 1966; 45: 144–51. 132

Authors' Biographies

AHMAD RASHAD SAAD

Ahmad Rashad Saad (BDS, MSc), is a researcher at the Tissue Engineering Laboratories, Faculty of Dentistry, Alexandria University, Egypt. He received his Bachelor of Dental Surgery (BDS) from the Faculty of Dentistry, Alexandria University, in 2006, and he is now a master student in the Materials Science Department, Institute of Graduate Studies and Research (IGSR), Alexandria University. The focus of his research is the preparation, characterization and use of biomaterials in oral-craniofacial tissue engineering and around dental implants.
dr.arashad@gmail.com

CHARLES SFEIR

Charles Sfeir (DDS, PhD), is an associate professor in the Departments of Oral Biology, periodontics, Bioengineering and the McGowan Institute for Regenerative Medicine. He received a Doctor of Dental Surgery (DDS) degree from the Université Louis Pasteur - Strasbourg France in 1990. He also has a degree in periodontology and holds a Ph.D. in molecular biology from Northwestern University. As the director of the Center for Craniofacial Regeneration, his research focuses on the development of bio-inspired materials for mineralized tissue engineering as well as understanding extracellular matrix proteins involved in mineralized tissue biology.
csfeir@dental.pitt.edu

DIMITRIOS TZIAFAS

Dimitrios Tziafas (DDS, PhD), is a professor and is the head of the Department of Endodontology, School of Dentistry, Aristotle University of Thessaloniki, Greece. He graduated as a dentist from the same school in 1977, undertaking the PhD in 1983. In 1987, he worked as a visiting post-doc researcher in the Institute Biologie Medicale, Faculty de Medicine, University Louis Pasteur, Strasbourg, France. He served as the national representative of Greece in the European Committee networks of Coordinated Science and Technology actions in the field of odontogenesis, oral facial development and regeneration and the president of IADR – Continental European Division until 2008. His research interests include cytodifferentiation mechanisms during tooth development and dental pulp repair, biology of mineralized tissues and pathogenesis of pulp inflammation.
dtziaf@dent.auth.gr

FATIMA N. SYED-PICARD

Fatima N. Syed-Picard (MS), is currently a bioengineering PhD student at the University of Pittsburgh. She received her Bachelor and Master degrees in Materials Science and Engineering at the University of Michigan. Her dissertation research involves using three-dimensional scaffold-less constructs engineered from dental pulp stem cells or bone marrow stromal cells as a model to study stem cell differentiation and the odontoblast and osteoblast microenvironments in Center for Craniofacial Regeneration, McGowan Institute for Regenerative Medicine.
fns4@pitt.edu

JEAN-CLAUDE PETIT

Jean-Claude Petit (BSc, LDS, MDent), is a professor and head of the Department of Oral Medicine and Periodontology at the University of the Witwatersrand, Johannesburg. He received the LDS degree in 1967 and MDent (1986) from the Universities of Louvain and Witwatersrand. His research interest is periodontal tissue regeneration.
Jean-Claude.Petit@wits.ac.za

JUNE TEARE

June Teare (MSc), is a senior scientist at the Bone Research Unit of the South African Medical Research Council and the University of the Witwatersrand Medical School, Johannesburg. She received her Master of Science degree from the University of the Witwatersrand in 2006. The focus of her research is the use of bone morphogenetic and transforming growth factor-β proteins in tissue engineering, particularly in the field of periodontal tissue regeneration.
june.teare@wits.ac.za

MANAL M. SAAD

Manal M. Saad (BDS, PhD), is currently a senior researcher in Tissue Engineering Laboratories, Alexandria University and lecturer of oral biology, Faculty of Dentistry, Pharos University- Alexandria, Egypt. She was graduated in 1984 from the Faculty of Dentistry, Alexandria University then received her PhD in Neuroscience from the Faculty of Medicine, Université de Montréal, Canada, in 1998. She presented her work in different national & international conferences related to biological evaluation of biomaterials in tissue engineering and around dental implants.
saad_manal@hotmail.com

MOHAMAD NAGEEB HASSAN

Mohamad Nageeb Hassan (BDS, MSc), is a researcher at the Tissue Engineering Laboratories, Faculty of Dentistry, Alexandria University, Egypt. He received his Bachelor of Dental Surgery (BDS) from the same faculty at 2006, and he is now a master student at the Materials Science Department, Institute of Graduate Studies and Research (IGSR) at the same university after having a diploma in materials science. The focus of his research is the preparation, characterization and use of biomaterials in oral-craniofacial bone engineering and around dental implants.
nageebdent@hotmail.com

MONA K. MAREI

Mona K. Marei (BDS, MScD, PhD), is a professor of prosthodontics, head of Tissue Engineering Laboratories, Faculty of Dentistry, Alexandria University, Egypt. She was graduated as a dentist from the School of Dentistry, Alexandria University in 1973, and received her postgraduate degree in major perio-prosthesis from Boston University in 1981. She is the founder of tissue engineering science and technology in Egypt since 1999. She has many international oral presentations and publications in the field of tissue engineering and regenerative dentistry. She is currently a board member of a number of national, African and international councils in addition to being an editorial board member in Artificial Organs, Tissue Engineering – journal part A, B and C and an associate editor in the Annals of Biomedical Engineering.
mona.marei@gmail.com

RANIA M. ELBACKLY

Rania M. Elbackly (BDS, MSc), is currently a PhD student in Laboratory of Regenerative Medicine, Department of Biology, Oncology and Genetics - University of Genoa - Italy. She was a member of the tissue engineering research team at the tissue engineering laboratories, Alexandria University, since 1999. In 2000, she graduated from the faculty of dentistry with honor degrees and went on to receive her masters in conservative dentistry in 2006. She has also presented at several national and international conferences including the world regenerate congress which was held in Pittsburgh, USA in 2006. She has several publications pertaining to the field of tissue engineering, and, in particular, to dental tissue engineering.
ranianoaman@gmail.com

SAMER H. ZAKY

Samer H. Zaky (BDS, PhD), is currently a postdoctoral associate-Center of Craniofacial Regeneration, McGowan Institute of Regenerative Medicine- University of Pittsburgh–USA. He received his Bachelor of Dental Surgery (BDS) in 2000 from Faculty of Dentistry, Alexandria University–Egypt, worked as a researcher in Tissue Engineering Laboratories for five years, and went on to receive his PhD in regenerative medicine in 2009 from the Laboratory of Regenerative Medicine, Department of Biology, Oncology and Genetics- University of Italy. The focus of his research is optimization of culture conditions for human bone marrow stromal cells proliferation and differentiation, in addition to Characterization of osteoconductive scaffolds to regenerate a critical-size bone defect together with the reestablishment of the stem cell niche.
shz33@pitt.edu

SAYURI YOSHIZAWA

Sayuri Yoshizawa (DDS, PhD), is currently a research assistant professor in the Department of Oral Biology and the Center for Craniofacial Regeneration at the University of Pittsburgh. She obtained her DDS and PhD from Okayama University's Graduate School of Medicine and Dentistry in Japan. Following her PhD, she pursued a post-doctoral training at the National Institute of Dental and Craniofacial Diseases under the guidance of Dr. Pamela Robey. Her research interests are engineering strategy of salivary glands and bone regeneration using biodegradable metal alloys.
say15@pitt.edu

SHINSUKE ONISHI

Shinsuke Onishi (DDS, PhD), received his dental training at the Nihon University School of Dental Medicine, Tokyo, Japan, and his oral surgery residency training at the Tokyo Women's Medical University Hospital. Dr. Onishi pursued specialty training in periodontology and Doctor of Philosophy (PhD) in biology at the State University of New York at Buffalo School of Dental Medicine (NY, USA) which he completed in 2009. He is currently conducting postdoctoral research in tissue engineering and regenerative medicine in the Center for Craniofacial Regeneration at the University of Pittsburgh, McGowan Institute for Regenerative Medicine.
sho12@pitt.edu

UGO RIPAMONTI

Ugo Ripamonti (MD, PhD), is a professor and director, Bone Research Laboratory, Research Unit of the South African Medical Research Council (MRC) and the University of the Witwatersrand, Johannesburg. He received his degree in medicine and surgery in 1978 *(cum laude)*, in 1985 he received his Master of Dentistry in Periodontics and Oral Medicine, and PhD in 1993. He has organized and was invited as a chairman and key note speaker for more than one hundred of international congresses and meetings. In addition, he has more than hundred and fifty of published papers in international journals including book chapters in bone and dental tissue engineering. His research interests are regenerative medicine, biomaterials, tissue induction and morphogenesis.
Ugo.ripamonti@wits.ac.za

Printed in the United States
by Baker & Taylor Publisher Services